S0-BWT-555

Developments in Economic Geology, 6

HYDROTHERMAL URANIUM DEPOSITS

Further titles in this series

1. *I.L. ELLIOTT and W.K. FLETCHER*
GEOCHEMICAL EXPLORATION 1974

2. *P.M.D. BRADSHAW*
CONCEPTUAL MODELS IN EXPLORATION GEOCHEMISTRY
The Canadian Cordillera and Canadian Shield

3. *G.J.S. GOVETT and M.H. GOVETT*
WORLD MINERAL SUPPLIES
Assessment and Perspective

4. *R.T. SHUEY*
SEMICONDUCTING ORE MINERALS

5. *J.S. SUMNER*
PRINCIPLES OF INDUCED POLARIZATION FOR GEOPHYSICAL EXPLORATION

Developments in Economic Geology, 6

HYDROTHERMAL URANIUM DEPOSITS

Robert A. Rich
Heinrich D. Holland
Ulrich Petersen

ELSEVIER SCIENTIFIC PUBLISHING COMPANY
Amsterdam — Oxford — New York 1977

ELSEVIER SCIENTIFIC PUBLISHING COMPANY
335 Jan van Galenstraat
P.O. Box 211, Amsterdam, The Netherlands

Distributors for the United States and Canada:

ELSEVIER NORTH-HOLLAND INC.
52, Vanderbilt Avenue
New York, N.Y. 10017

Library of Congress Cataloging in Publication Data

Rich, Robert A
Hydrothermal uranium deposits.

(Developments in economic geology ; 6)
A revision of ERDA report GJO-1640 prepared by R.A. Rich and H.D. Holland, published in 1975.
Includes bibliographies and indexes.
1. Uranium ores. 2. Hydrothermal deposits. I. Holland, Heinrich D., joint author. II. Petersen, Ulrich, 1927- joint author. III. Title. IV. Series.
TN490.U7R47 1976 553'.493 77-1411
ISBN 0-444-41551-3

ISBN: 0-444-41551-3

Printed in The Netherlands

Red hematitic alteration associated with pitchblende vein zone, 20th level, Fay Mine, Beaverlodge district, Saskatchewan; painted white lines mark the boundaries of the ore zone (photo by U. Petersen).

This book is dedicated to the
memory of Armin Petersen
1954—1977

"I lift up mine eyes unto
the hills"
Psalm 121

TABLE OF CONTENTS

PART II — DESCRIPTIONS OF HYDROTHERMAL URANIUM DEPOSITS

LIST OF FIGURES

FIGURE

LIST OF TABLES

TABLE

Preface

The last few years have seen a remarkable renaissance of interest in uranium and its ores. The reasons for this are obvious. Our supplies of petroleum and natural gas have rather suddenly become uncertain and are rapidly dwindling. The prospect of a rapid increase in coal mining and burning is clouded by health and environmental hazards. Solar power is still of marginal use as a major direct source of energy, and the future of nuclear fusion is uncertain. It is likely therefore that uranium-fueled fission reactors will be needed to provide an important share of the world's energy during the next fifty years; as a consequence there is considerable concern about the sufficiency of reasonably inexpensive uranium ore which can be mined without serious environmental consequences.

Most of the uranium mined in the United States has come from "Colorado Plateau" and "Wyoming roll front" sandstone-type depositsm Hydrothermal uranium deposits, i.e. deposits which occur as discordant veins, stockworks, breccia zones, and irregular bodies of metasomatized rock, account for only about 4% of the total U.S. production plus reserves, but these deposits have been major and even dominant sources of uranium in Canada, Europe, Australia and Zaire. There are good reasons to believe that additional prospecting for hydrothermal uranium deposits will be successful in the United States and elsewhere, and that the chances for success will be enhanced by a better understanding of the properties, distribution and origin of this deposit type. Our book is a first attempt toward such an understanding. It is an outgrowth of a report prepared in 1973 by one of us (U.P.) for the Exxon Corporation. In 1975 this report was rewritten and expanded by two of us (R.A.R. and H.D.H.) as part of work done for the U.S. Energy Research and Development Administration (ERDA) under Contract #AT(05-1)-1640. Our final report was placed on open file as ERDA Report GJO-1640 late in 1975. The text of *Hydrothermal Uranium Deposits* evolved as a revision of the ERDA report by all three authors early in 1976.

The book is divided into two parts. The first part deals with hydrothermal uranium deposits as a whole. We have attempted to summarize their important properties, to describe their position in the geochemical cycle of uranium, and to develop a genetic model that can explain their occurrence and their mineralogy. In the course of our literature search and field work we have become impressed with the geochemical similarities between hydrothermal uranium deposits and uranium deposits in sandstones. It seems likely that in the formation of both types of deposits uranium is transported in oxidizing solutions and deposited in response to its reduction from the hexavalent to the tetravalent state. The implications of this hypothesis are potentially very important in the search for new hydrothermal uranium deposits. The second part of the book is devoted to descriptions of many of the major hydrothermal uranium deposits of the world. We have attempted to assemble all of the pertinent recent literature and to summarize the salient features of the deposits, particularly those that bear on their origin.

The authors owe a debt of gratitude to a large number of people and organizations for help in the preparation of this book. The Exxon Corporation kindly consented to the use of their 1973 report, and ERDA gave permission for the publication of our revised open file report. We particularly wish to express our gratitude to Dr. Hans Adler of ERDA and Dr. Bernard Poty of the Centre de Recherches Pétrographiques et Géochimiques, Nancy, France, for their kind help at many points during the preparation of this book, and to Robert F. Quirk for the preparation of the Index. Special thanks are due to Dr. Poty and Dr. J. Geffroy for organizing and leading a field trip for two of us (H.D.H. and U.P.) to the Bois Noirs-Limouzat, Margnac and Fanay uranium mines in the French Massif Central. We also wish to thank Dr. J.T. Nash of the U.S. Geological Survey for his help in planning visits to several hydrothermal uranium deposits in the western United States. Nash's field trip to the Midnite mine was especially appreciated. One of us (HDH) wishes to thank the Guggenheim Foundation for support during the preparation of the manuscript.

We further acknowledge the generous help of E. Bruner, J. Haley and J. Cohen of the Cotter Corporation for the Schwartzwalder mine visit; I.W. Mathisen, J.R. Andrus and D. Robbins of Western Nuclear, Inc. for visits to the Marysvale district, Utah, and the Mount Spokane area, Washington; E. Craig and N. Lehrman of Dawn Mining Company for the visit to the Midnite mine; Dr. E.E.N. Smith, D. Ward and F. Gentile of Eldorado Nuclear Ltd. for the Fay mine visit; and C.R. Burkhart, W. MacDonald and R. Janes of Gulf Minerals Canada Ltd. for our visit to the Rabbit Lake mine.

PART I

GEOLOGY, GEOCHEMISTRY AND ORIGIN OF HYDROTHERMAL URANIUM DEPOSITS

Chapter 1
THE DISTRIBUTION OF URANIUM

1. General statement

Uranium is a lithophile element which is enriched in the upper crust (Dybek, 1962; Clark et al., 1966; Rogers and Adams, 1967). The average crustal abundance of uranium is about 2 ppm. In nature uranium occurs in the tetravalent (U^{+4}) and in the hexavalent (U^{+6}) state; in most subsurface environments uranium is present as U^{+4}, because the U^{+6} state is stable only under oxidizing conditions. The combination of the relatively large size and high charge of U^{+4} prevents it from entering the crystal lattices of the major rock-forming minerals except in trace amounts; however, significant quantities of uranium are frequently found in accessory minerals such as thorianite, thorite, thorogummite, allanite, xenotime, zircon, fluorite, apatite, and barite. Much of the uranium in rocks is only loosely held; this fraction of the uranium probably occurs in films coating grains of rock-forming minerals, and can be easily leached by dilute acids (Larsen et al., 1956; Barbier et al., 1967; Szalay and Samsoni, 1969). U^{+4} is readily converted to U^{+6} under oxidizing conditions. Leaching of uranium from weathered outcrops is rapid (Barbier and Ranchin, 1969), and is enhanced by the formation of complex uranyl ions with carbonate and sulfate ions (Katz and Rabinowich, 1951; Dybek, 1962; Garrels and Larsen, 1959;). Uranium can be removed efficiently from oxidizing solutions by reduction. In nature reduction is most frequently due to the presence of sulfide, ferrous iron, carbonaceous matter, or hydrocarbons. Uranium can also be removed from aqueous solutions by adsorption on clays, organic matter, and iron hydroxides.

Uraninite is the most abundant uranium mineral and the only commonly occurring U^{+4} mineral. Coffinite is found in minor amounts in some hydrothermal uranium deposits. There are, however, many minerals containing hexavalent uranium. Frondel (1958) and Frondel et al. (1967) have summarized the mineralogy of uranium. Pitchblende, the fine-grained, massive, botryoidal, or sooty variety of uraninite, is the most important uranium ore mineral of hydrothermal deposits. The ideal formula for pitchblende is UO_2, but in nature some U^{+6} is always present. $(U^{+4}_{1-x}, U^{+6}_{x})\,O_{2+x}$ is therefore a more realistic formula for pitchblende. Natural pitchblendes exhibit compositions ranging up to $UO_{2.6}$ (Frondel et al., 1967). It is not known at present whether the compositional variation of pitchblende reflects differences in the conditions of primary deposition or the operation of post-depositional oxidation processes. Magmatic and pegmatitic uraninite usually contains large amounts of thorium and rare earths, but hydrothermal pitchblendes do not (Frondel et al., 1967).

2. Igneous rocks

Table 1-1 shows that uranium tends to be enriched in the late stage members of igneous differentiation series (Larsen et al., 1956). High uranium concentrations usually correlate well with high concentrations of potassium and silica. Granitic and alkalic rocks usually contain 2-4 ppm uranium, but felsic rocks (and especially their pegmatitic derivatives) containing more than 10 ppm uranium are not uncommon. Bostonite dikes from the Central City district, Colorado, for example, contain on the order of 100 ppm uranium (Phair, 1952). Mafic and ultramafic rocks generally contain less than 1 ppm uranium.

Table 1-1. Uranium Content (in ppm) of Igneous Rock Types (Data from compilations by Clark et al., 1966; Rogers and Adams, 1967)

Rock Type	Average	Range
Dunites, olivine nodules, and peridotite inclusions	-	0.003-0.05
Pyroxenites	0.70	-
Eclogites (crustal, metamorphic, kimberlitic, and basaltic inclusions)	-	0.013-0.80
Mafic igneous rocks	0.9	<0.2-3.4
Diorites and quartz diorites	2.0	<0.5-11.5
Granodiorites	2.6	<1-9
Granitic rocks	-	2.2-15
Silicic igneous rocks	4.7	<1-21
Silicic extrusive rocks (rhyolites and dacites)	5.0	-
Alkalic intrusive rocks	-	0.04-19.7

Felsic minerals in igneous rocks usually contain much less uranium than biotite, hornblende and pyroxene. The highest uranium concentrations are found in accessory minerals such as xenotime, zircon, monazite, sphene, allanite, epidote, and apatite. The uranium content reported for various igneous rock-forming minerals are given in Table 1-2. Many of the rather high values in common rock-forming minerals are probably the result of solid inclusions of uranium-rich minerals (e.g. zircons in biotite). It is obvious from the data presented in Table 1-2 that the tenor of uranium in a rock can be greatly affected by the presence of only small amounts of accessory minerals. This is especially true where one or more uranium minerals such as uraninite, thorite, thorianite, brannerite, and euxenite occur as accessories.

Table 1-2. Uranium Content (in ppm) of Igneous Rock-Forming Minerals (data from compilations by Clark et al., 1966; Rogers and Adams, 1967).

Major Minerals:	Range	Accessory Minerals:	Range
Quartz	0.1-10	Allanite	30-1,000
Feldspar	0.1-10	Apatite	5-150
Muscovite	2-8	Epidote	20-200
Biotite	1-60	Garnet	6-30
Hornblende	0.2-60	Ilmenite	1-50
Pyroxene	0.01-50	Magnetite	1-30
Olivine	~0.05	Monazite	500-3,000
		Sphene	10-700
		Xenotime	300-35,000
		Zircon	100-6,000

In France it has been clearly established that hydrothermal uranium deposits are closely associated with uranium-rich ("fertile") two-mica granites (e.g. see Gangloff, 1970). Similarly, Marjaniemi and Basler (1972) have found that uranium concentrations are abnormally high in individual Cenozoic and Mesozoic plutons spatially associated with known hydrothermal uranium deposits of the western United States. A list of references to data on the uranium content of granitic rocks is given in Appendix I.

3. Sedimentary rocks

The distribution of uranium in sedimentary rocks is summarized in Table 1-3. Average sedimentary rock has about the same uranium content as average igneous rock. Among the different sedimentary rocks, however, the distribution of uranium is quite variable. Clean sandstones are usually rather poor in uranium, and the uranium which is present is largely concentrated in "heavy" accessory minerals. Pure limestones are similarly poor in uranium. However, uranium enrichment in sedimentary rocks can be pronounced, particularly in shales and phosphorites. The uranium content of shales correlates with clay content and with the concentration of organic carbon. Slowly deposited carbonaceous shales, such as the Late Devonian-Early Mississippian Chattanooga shale of the south-central United States (averaging 70 ppm U_3O_8) and the Cambrian Alum shale of Sweden (averaging 300 ppm U_3O_8), are potential uranium ores. Organic matter in marine sediments undoubtedly acts as a reducing agent for U^{+6} in sea water, but the exact functional relation between the reduced carbon and uranium contents of shales remains unclear. Uranium is coprecipitated readily with apatite; it can therefore be strongly enriched in marine phosphorites. In the southern United States uranium is recovered as a by-product of phosphorite mining from rock containing 100-200 ppm U_3O_8.

Table 1-3. Uranium Content (in ppm) of Sedimentary Rock Types (Data from compilations by Clark et al., 1966; Rogers and Adams, 1967).

	Average	Range
Fine-grained Clastics:		
Common Shales	3.7	1-13
North American gray and green shales	3.2	1.2-12
Mancos shale (western U.S.A.)	3.7	0.9-12
Black shales	-	3-1250
Coarse-grained Clastics:		
Sandstones (including arkoses and graywackes)	-	0.45-3.2
Orthoquartzites	0.45	0.2-0.6
Carbonates:		
Carbonate rocks	2.2	0.1-9
Russian carbonates	2.1	-
North American carbonates	2.2	0.65-8.8
California limestones	1.3	0.03-4.9
Florida limestones	2	0.5-6
Other Sedimentary Rocks:		
Marine phosphorites	-	50-300
Evaporites	-	0.01-0.43
Bentonites	5.0	1-21
Bauxites	8.0	3-27

4. Metamorphic rocks

The uranium content of metamorphic rocks is quite variable, and tends to reflect the uranium concentration of their protoliths. However, some very high grade metamorphic rocks are apparently depleted in uranium relative to their lower grade and unmetamorphosed equivalents, and it is possible that uranium moves upward in the crust during granulite grade metamorphism (e.g. see Heier and Adams, 1965).

5. Natural waters

Most natural waters contain no more than a few ppb dissolved uranium. The concentration of uranium in sea water has a relatively constant value of 1-4 ppb. The uranium content of most surface and near surface continental waters is somewhat lower, and as a class they are much more variable; values ranging from less than 0.1 ppb to more than 1000 ppb have been reported in the literature. In natural waters, however, uranium concentrations greater than 100 ppb are quite rare, and have generally been found only in aquifers containing uranium mineralization. This observation has obvious implications for hydrogeochemical prospecting. References to data on the uranium content of continental ground waters are listed in Appendix II.

The dissolved uranium content of continental ground waters is a complex function of numerous variables; among these, aquifer rock type and ground water composition (especially f_{O_2}) are probably most important (Barker and Scott, 1958; Scott and Barker, 1962; Lopatkina, 1964). Table 1-4 summarizes the data of Scott and Barker (1962) on the relation of country rock type to the uranium content of dilute, low temperature ground waters in the United States. These data suggest that a correlation exists between the average uranium content of common sedimentary and igneous rocks (Tables 1-1 and 1-3) and the average uranium concentration of contained ground waters. Siltstones and shales and silicic igneous rocks are uranium-rich rock types, and these rocks also tend to have ground waters with higher than average dissolved uranium concentrations. 43% of the analyzed waters from siltstones and shales and 36% of the analyzed waters from silicic igneous rocks have anomalous (> 4 ppb) uranium concentrations. The contrast between the uranium content of waters from uranium-rich silicic and uranium-poor basic igneous rocks is particularly striking. Metamorphic rocks and sandstones and conglomerates also frequently contain uranium-rich ground waters. Because the uranium content of a metamorphic rock is usually similar to that of its protolith, it is not surprising that some metamorphic terranes (e.g. metapelites and silicic metavolcanics) contain uranium-rich ground waters. The occurrence of uranium-rich waters in sandstones and conglomerates, however, would not be predicted on the basis of the average uranium content of these rocks. It is probable that the occurrence of many uranium-rich ground waters in sandstones and conglomerates is related to the leaching of zones of sandstone-type uranium mineralization. Indeed, all three analyzed groundwaters containing extraordinarily high (> 100 ppb) uranium concentrations are from sandstone aquifers, and the two highest analyses are from areas of known uranium mineralization.

Table 1-4. Uranium Content of Dilute, Low Temperature Ground Waters as a Function of Terrane Rock Type (data from Scott and Barker, 1962).

Terrane	No. of Samples	U content in ppb Range	U content in ppb Average	No. of samples with > 4 ppb	% of samples with > 4 ppb
Igneous:					
Silicic	33	0-32	4.5	12	36
Basic and intermediate	18	0-9.2	0.9	1	6
Sedimentary:					
Sandstone and conglomerate	132	0-2100	26.2 (2.2*)	22	17
Siltstone and shale	14	0-69	10.6	6	43
Limestone and dolomite	89	0-33	2.0	11	12
Sand and gravel	87	0-74	2.5	13	15
Metamorphic:					
Undifferentiated	34	0-37	4.4	8	24
Totals:	407	0-2100	10.6 (2.8*)	73	18

* Eliminating 3 analyses of waters containing more than 100 ppb U.

References for Chapter 1

Barbier, J. and Ranchin, G., 1969, Géochimie de l'uranium dans le Massif de Saint-Sylvestre (Limousin-Massif Central Français). Occurrences de l'uranium géochimique primaire et processus de remaniements: Sci. Terre Mém. 15, 115-157.

Barbier, J., Carrat, H.G. and Ranchin, G., 1967, Présence d'uraninite en tant que minéral accessoire usuel dans les granites à deux micas uranifères du Limousin et de la Vendée: Acad. Sci. Comptes Rendus, Ser. D., *264,* 2436-2439.

Barker, F.B. and Scott, R.C., 1958, Uranium and radium in the ground water of the Llano Estacado, Texas and New Mexico: Am. Geophys. Union Trans., *39,* 459-466.

Clark, S.P., Jr., Peterman, Z.E. and Heier, K.S., 1966, Abundances of uranium, thorium, and potassium: in Clark, S.P., Jr., ed., *Handbook of Physical Constants,* Revised edition, Geol. Soc. America Mem. 97, 521-541.

Dybek, I.J., 1962, Zur Geochemie und Lagerstättenkunde des Urans: Clausthaler Hefte zur Lagerstättenkunde und Geochemie der mineralischen Rohstoffe, Heft 1, 163 pp.

Frondel, C., 1958, Systematic mineralogy of uranium and thorium: U.S. Geol. Survey Bull. 1064, 400 pp.

Frondel, J.W., Fleischer, M. and Jones, R.S., 1967, Glossary of uranium- and thorium-bearing minerals, 4th edition: U.S. Geol. Survey Bull. 1250, 69 pp.

Gangloff, A., 1970, Notes sommaires sur la géologie des principaux districts uranifères étudiés par la C.E.A.: in *Uranium Exploration Geology,* 77-105, Internat. Atomic Energy Agency, Vienna.

Garrels, R.M. and Larsen, E.S., 3rd, 1959, Geochemistry and mineralogy of the Colorado Plateau uranium ores: U.S. Geol. Survey Prof. Paper 320, 236 pp.

Heier, K.S. and Adams, J.A.S., 1965, Concentration of radioactive elements in deep crustal material: Geochim. et Cosmochim. Acta, *29,* 53-61.

Katz, J.J. and Rabinowitch, E., 1951, *The Chemistry of Uranium,* Part I: National Nuclear Energy Series, McGraw Hill, 609 pp.

Larsen, E.S., Jr., Phair, G., Gottfried, D. and Smith, W.S., 1956, Uranium in magmatic differentiation: Internat. Conf. Peaceful Uses of Atomic Energy, *6,* 240-247, United Nations.

Lopatkina, A.P., 1964, Characteristics of migration of uranium in the natural waters of humid regions and their use in the determination of the geochemical background for uranium: Geochem. International, 788-795.

Marjaniemi, D.K. and Basler, A.L., 1972, Geochemical investigations of plutonic rocks in the western United States for the purpose of determining favorability for vein-type uranium deposits: U.S. Atomic Energy Comm. Rept. GJO-912-16, 134 pp.

Phair, G., 1952, Radioactive Tertiary porphyries in the Central City district, Colorado, and their bearing upon pitchblende deposition: U.S. Atomic Energy Comm. Rept. TEI-247, 53 pp.

Rogers, J.J.W. and Adams, J.A.S., 1967, Uranium: in Wedepohl, K.H., ed., *Handbook of Geochemistry, v. 2,* pt. 1, Chap. 92, 50 pp., Springer-Verlag, Berlin.

Scott, R.C. and Barker, F.B., 1962, Data on uranium and radium in ground water in the United States: U.S. Geol. Survey Prof. Paper 426, 115 pp.

Szalay, S. and Samsoni, Z., 1969, Investigation of the leaching of uranium from crushed magmatic rock: Geochem. International, *6,* 613-623.

Chapter 2
CHARACTERISTICS OF HYDROTHERMAL URANIUM DEPOSITS

1. Geological setting

Hydrothermal uranium deposits occur in diverse geological environments and in a wide variety of rock types ranging in age from Precambrian to Tertiary. Virtually all types of igneous, sedimentary, and metamorphic rocks have been found to host hydrothermal uranium mineralization. Even within a single district host rocks may vary widely with respect to mineralogy, composition, texture, and structural competence. In spite of this great diversity, some generalizations do seem to be valid:

1. Most host rocks for hydrothermal uranium deposits in the United States are Precambrian, late Mesozoic, or Tertiary in age. In the rest of the world, host rocks are predominantly Precambrian and late Paleozoic.
2. Host rocks in most hydrothermal uranium deposits are competent felsic igneous and metamorphic rocks. Petersen (unpublished report, 1973) noted a strong correlation between the occurrence of hydrothermal uranium deposits and granitic rocks. Using the data of Walker and Osterwald (1963) for 548 hydrothermal uranium occurrences in the United States, Petersen found that no less than 83% of these deposits occurred in or closely associated with pre-Tertiary granitic rocks and Tertiary intrusives. As mentioned in Chapter 1, there is a strong positive correlation between the distribution of pitchblende veins and granites containing anomalously large amounts of uranium (>5 ppm).
3. Non-igneous host rocks usually contain carbonaceous matter, clays, ferrous silicates, sulfides, or phosphates.

Hydrothermal uranium deposits are epithermal or mesothermal in character; the introduced minerals fill open spaces created by structural deformation. Uranium veins usually occupy faults, joints, or fracture zones. Uranium-bearing stockworks, pipe-like metasomatic replacement bodies, and mineralized breccias are also found. Hydrothermal uranium veins frequently exhibit cataclastic textures. Individual pitchblende veins are usually small, but some are as wide as a few meters and may extend along strike or down dip for hundreds of meters. Such veins sometimes combine to form systems which extend for a kilometer or more. Most hydrothermal uranium deposits are quite shallow, rarely extending to depths greater than 300 m; however, mining at the Schwartzwalder mine, Colorado, and in the Beaverlodge district, Saskatchewan, has already proceeded to depths of 700 and 1,500 m respectively. Most hydrothermal uranium deposits contain only a few tons to a few million tons of ore. A few large deposits contain more than 10,000 tons of uranium. Average mined grade ranges from 0.10 to 1.0 weight $\% U_3O_8$.

2. Age of deposits

The age of the known hydrothermal uranium deposits ranges from middle Precambrian to late Tertiary. The absence of hydrothermal uranium deposits of Archean age (>2.5 b.y.) is striking (see Smith, 1974, p. 525). In the United States the major deposits have late Mesozoic to Tertiary ages. Pitchblende ages from a single deposit sometimes show a broad range due to the presence of more than one generation of ore. Such deposits are thought to reflect the apparent ease with which pitchblende can be mobilized and redeposited. The ages of many individual hydrothermal uranium deposits are included in Part II of this book.

3. Mineralogy

The mineralogy of hydrothermal uranium deposits is generally simple. Deposits with complex paragenesis (e.g. the cobalt-nickel arsenide type) usually have a mineralogically simple uranium stage. In virtually all deposits pitchblende is the only important hypogene uranium mineral. Hydrothermal pitchblende may be massive, botryoidal, or sooty. Idiomorphic UO_2 (uraninite) is quite rare in hydrothermal veins, but small uraninite crystals have been found at Shinkolobwe and in several other deposits. Natural pitchblendes vary markedly in their degree of oxidation and hydration.

In some hydrothermal uranium deposits with simple mineralogy pitchblende is the only mineral present, but most veins contain at least some gangue, usually quartz or calcite, and small amounts of sulfides. Pyrite and marcasite are particularly common. Hematite is present either as a wall rock alteration or vein mineral in most hydrothermal uranium deposits (see Tables 2-1, 2-3 and 2-4). A study of paragenetic diagrams for 41 of the uranium deposits listed in Table 2-2 showed the presence of the following minerals in more than half of the deposits: pyrite/marcasite (100%), base metal sulfides (88%), quartz (81%), hematite (71%), and carbonate minerals (55%). The complement of minerals accompanying pitchblende varies from district to district and sometimes from vein to vein within a single district.

Hydrothermal uranium deposits with complex mineralogy generally contain a variety of sulfides and sulfosalts. Sulfides and especially sulfarsenides and arsenides of nickel and cobalt are often present; however, it should be noted that these minerals have almost always been deposited during a later stage than pitchblende (see Table 2-3). Significant amounts of silver and bismuth minerals often accompany the arsenides. The presence or absence of large amounts of base metals in general (and the cobalt-nickel-silver-arsenic-bismuth assemblage in particular) could reflect the presence or absence of a source for the metals rather than a fundamental difference in the conditions of transport and deposition. In complex deposits fluorite, barite, and one or more carbonate minerals often accompany quartz as major gangue minerals. Everhart and Wright (1953) suggested a correlation between host rock composition and gangue mineralogy. For example, deposits in metamorphic terranes often have dominantly carbonate gangue, whereas deposits in granitic intrusive rocks generally have siliceous gangue.

Gangue minerals of hydrothermal uranium deposits sometimes show distinctive effects resulting from radiation damage. The usual effects are marked changes in mineral thermoluminescence and color, and reduction or loss of crystallinity (metamictization). Common radiation effects include the development of smoky color in quartz, deep purple to black color in fluorite, and pleochroic halos in biotite. Walker and

Adams (1963) suggested the possibility that the hematitic alteration associated with many uranium veins may be the result of radiation-induced oxidation of Fe^{+2}, but this seems unlikely.

4. Paragenesis

Detailed paragenetic studies have only been reported for a relatively small number of uranium deposits. Only fragmentary data are available for most deposits, and reliable data regarding minerals which were precipitated together with pitchblende are rare.

Hydrothermal uranium deposits for which paragenetic diagrams have been published are listed in Table 2-2; copies of the published paragenetic diagrams are included with the description of the deposits in Part II of this book. A statistical analysis was made of the vein mineralogy of only those uranium deposits for which published paragenetic diagrams are available, because most written descriptions are incomplete. Unfortunately, this approach excluded some important hydrothermal uranium deposits and included many minor producers and non-producers. For example, small French deposits of simple paragenesis are over-represented in the group of 41 deposits considered.

In spite of the shortcomings of the available data, several useful observations can be made on the basis of the paragenetic data summarized in Tables 2-3 and 2-4:

1. Pitchblende precipitated alone at some point in the depositional sequence of 58% of the deposits studied.
2. Where pitchblende was reported to have been deposited simultaneously with one or more other minerals, only the following assemblages were reported for more than 20% of the deposits: pitchblende + pyrite/marcasite (32% of the deposits), pitchblende + quartz (24%), pitchblende + pyrite/marcasite + quartz (22%).
3. Minerals commonly deposited at the same time as pitchblende are: pyrite/marcasite (66% of the deposits), quartz (51%), base metal sulfides (36%), carbonate (24%), and hematite (24%). All other minerals were deposited together with pitchblende in less than 10% of the 41 deposits considered. Consequently, these minerals are not thought to be significant indicators of the conditions prevailinig during pitchblende deposition. Fluorite, barite, and the cobalt-nickel arsenides are absent from the group of minerals which are commonly deposited simultaneously with pitchblende.
4. All 41 deposits contain pyrite/marcasite, although not always in the stage of pitchblende deposition.
5. Most deposits contain base metal sulfides (90% of the deposits), quartz (83%), hematite (76%), and carbonate (51%).
6. Pyrite/marcasite and hematite were precipitated simultaneously in 29% of the deposits studied. They precipitated together with pitchblende in 24% of the deposits.
7. Fluorite occurs in 34% and barite in 20% of the deposits considered, but not in the stage of uranium deposition.
8. In 90% of the deposits the first appearance of pitchblende in the paragenetic sequence is early. However, one or more generations of pitchblende can occur. These can appear anywhere in the paragenetic sequence following the main pitchblende stage.
9. In 77% of the deposits reporting the presence of hypogene hematite, pitchblende is first deposited with or after the initial deposition of hematite.

Table 2-1. Characteristics of Some Hydrothermal Uranium Deposits (data from Part II of this book).

Location	Deposit(s)	Granite Association	Red bed (or clean sandstone) Association	Co-Ni Arsenides	Selenide(s)	Hematite	Carbonate(s)	Fluorite	Barite
Canada									
Saskatchewan	Beaverlodge district	X	X	X	X	X	X	X	X
	Rabbit Lake mine	(X)	X	(X)		(X)	X		
	Cluff Lake mine	(X)	X	X	X	X	X		
Northwest Territories	Great Bear Lake district	X	X	X		X	X	X	X
	Rayrock mine	X				X			
Labrador	Makkovik-Seal Lake area	X	(X)			X	X	X	
United States									
Arizona	Orphan mine		X	X		X	X		X
Colorado	Front Range district	X	X	X		X	X	X	X
	Marshall Pass district	X				X			
	Cochetopa district	X	X						X
Idaho	Sunshine mine	X		X		X	X		X
Nevada	Reese River district	X							
Oregon	Lakeview district	X							
Utah	Marysvale district	X				X	X	X	X
Washington	Midnite mine	X				(X)	X		
Australia									
Darwin Region, N.T.	Rum Jungle district	X	(X)	X		X	X	X	
	S. Alligator River district	X	(X)	X	X	X	X		
	Alligator Rivers district	X							
Queensland	Mary Kathleen mine	X					X	X	
Central Europe	Erzgebirge region	X	(X)	X	X	X	X	X	X
Czechoslovakia	Pribram district	X	(X)	X		X	X		X
	Labe Lineament region	X				X	X		
France	Massif Central region	X	(X)			X	X	X	X
Great Britain	Cornwall district	X	X	X		X	X	X	X
Portugal	Urgeirica district	X				X	X		
	Pinhel deposit	X							
Gabon	Mounana deposit	X	X				X		X
	Boyindzi deposit	X	X						
	Oklo deposit	X	X						
South West Africa	Rössing deposit	X				X		X	
Zaire	Shinkolobwe deposit	(X)		X*	X		X		
	Swambo deposit	(X)		X*	X	X	X		
	Kalongwe deposit	(X)		X*			X		

*Co-Ni sulfides only

Legend
X = present.
(X) = possibly present.

Table 2-2. Hydrothermal Uranium Deposits and Districts with Published Paragenetic Diagrams.

Country	District or Region	Deposit	Reference
Canada	Beaverlodge district	-	Robinson (1955)
		Fay mine/Bolger pit	Sassano et al. (1972)
		Martin Lake mine	Smith (1952)
	Great Bear Lake district	-	Badham et al. (1972)
		Eldorado mine	Ruzicka (1971)
		Echo Bay mine	Robinson and Ohmoto (1973)
		Terra mine	Robinson and Badham (1974)
United States			
Arizona	Grand Canyon region	Orphan mine	Gornitz and Kerr (1970)
Colorado	Ralston Buttes district	Schwartzwalder mine	Walker and Adams (1963)
		Union Pacific prospect	" "
	Central City district	-	" "
		Wood and E. Calhoun mines	Drake (1957)
	Marshall Pass district	Lookout claim	Gross (1965)
Michigan	Baraga County	Huron River deposit	Walker and Adams (1963)
	Gwinn district	Francis mine	" "
	Iron River district	Buck mine	" "
		Sherwood mine	" "
Utah	Marysvale district	-	" "
Central Europe	Erzgebirge region	-	Naumov et al. (1971)
		Jáchymov district	Ruzicka (1971)
	Western Erzgebirge region	-	"
France	Limousin region	Henriette	Geffroy & Sarcia (1960)
		La Besse	Cariou (1964)
		Le Brugeaud	Geffroy & Sarcia (1960)
		Margnac	" "
		Sagnes	" "
		Sapinière	Geffroy & Sarcia (1955)
		Villard	Geffroy & Sarcia (1960)
	Forez region	Bigay	" "
		Bois des Fayes	" "
		Viaduc-des-Peux	Geffroy & Sarcia (1955)
		Bois Noirs-Limouzat	Cuney (1974)
		Saint Priest	Geffroy & Sarcia (1960)
		Saint Rémy	Geffroy & Sarcia (1955)
	Morvan region	Bauzot	Carrat (1962)
		Huis Jacques	"
		La Faye	"
		Les Brosses	"
		Les Jalerys	"
		Les Vernays	"
	Vendée region	Les Ruaux	Geffroy & Sarcia (1960)
		Ecarpière	" "
		La Chapelle Largeau	" "
		La Commanderie	" "
	Miscellaneous deposits	Kruth	" "
		Guern	Germain et al. (1964)
		Kerségalec	Geffroy & Sarcia (1960)
		Quistiave	" "
		Cellier	Cariou (1964)
		Pierres Plantées	" "
U.S.S.R.	(An unidentified arsenide-bearing uranium deposit)		Ruzicka (1971)

Table 2-3. Mineralogy and Paragenesis of 41 Hydrothermal Uranium Deposits (Data from the paragenetic diagrams for the individual deposits listed in Table 2-2).

	35 Deposits without Co-Ni Arsenides			6 Deposits with Co-Ni Arsenides			All 41 Deposits		
	# of deposits	%		# of deposits	%		# of deposits	%	
Mineralogy:									
Base metal sulfides	31	89		6	100		37	90	
Quartz	28	80		6	100		34	83	
Hematite	27	77		4	67		31	76	
Carbonate	16	46		5	83		21	51	
Fluorite	12	34		2	33		14	34	
Barite	5	14		3	50		8	20	
Paragenesis*									
1) Pitchblende-early	33	94		4	67		37	90	
Pitchblende-intermediate	2	6		2	33		4	10	
	All 35 deposits		27 hematite-bearing deposits	All 6 deposits		4 hematite-bearing deposits	All 41 deposits		31 hematite-bearing deposits
	# of deposits	%	%	# of deposits	%	%	# of deposits	%	%
2) Pitchblende with or after hematite	21	60	78	3	50	75	24	59	77
Pitchblende before hematite	6	17	22	1	17	25	7	17	23

Table 2-3 continued.

			28 quartz-bearing deposits			6 quartz-bearing deposits			34 quartz-bearing deposits
			%			%			%
3) Pitchblende after quartz	23	66	82	5	83	83	28	68	82
Pitchblende with quartz	3	8	11	1	17	17	4	10	12
Pitchblende before quartz	2	6	7	0	0	0	2	5	6
4) Pitchblende before pyrite/marcasite	21	60		2	33		23	56	
Pitchblende after pyrite/marcasite	10	28		3	50		13	32	
Pitchblende with pyrite/marcasite	4	11		1	17		5	12	
5) Pitchblende before arsenides	-	-		5	83		-	-	
Pitchblende after arsenides	-	-		1	17		-	-	
			16 carbonate-bearing deposits			5 carbonate-bearing deposits			21 carbonate-bearing deposits
			%			%			%
6) Pitchblende before carbonate	12	34	75	2	33	40	14	34	67
Pitchblende after carbonate	4	11	25	3	50	60	7	17	33
			31 deposits with base metal sulfides			6 deposits with base metal sulfides			37 deposits with base metal sulfides
			%			%			%
7) Pitchblende before base metal sulfides	28	80	90	5	83	83	33	80	89
Pitchblende with base metal sulfides	2	6	6	0	0	0	2	5	5
Pitchblende after base metal sulfides	1	3	3	1	17	17	2	5	5

*Based on the first appearance of minerals.

Table 2-4. Frequency of Pitchblende Assemblages and Co-deposition Pairs for 41 Hydrothermal Uranium Deposits (data from published paragenetic diagrams included in Part II of this book).

		35 Deposits without Arsenides		6 Deposits with Arsenides		All 41 Deposits	
	Assemblage	# of deposits	%	# of deposits	%	# of deposits	%
One phase:	1. pbl	21	60	3	50	24	58
Two phase:	2. pbl + py/mc	12	34	1	17	13	32
	3. pbl + qtz	8	23	2	33	10	24
	4. pbl + b.m.s.	5	14	0	0	5	12
	5. pbl + carb	2	6	0	0	2	5
	6. pbl + hem	1	3	1	17	2	5
	7. pbl + fluor	1	3	0	0	1	2
	8.pbl + n.e.	0	0	1	17	1	2
Three phase:	9. pbl + qtz + py/mc	8	23	1	17	9	22
	10. pbl + carb + hem	3	8	0	0	3	7
	11. pbl + qtz + carb	2	6	0	0	2	5
	12. pbl + qtz + hem	2	6	0	0	2	5
	13. pbl +qtz + b.m.s.	2	6	0	0	2	5
	14. pbl + carb + py/mc	2	6	0	0	2	5
	15. pbl + carb + b.m.s.	2	6	0	0	2	5
	16. pbl + py/mc + b.m.s.	2	6	0	0	2	5
	17. pbl + qtz + fluor	1	3	0	0	1	2
	18. pbl + py/mc + fluor	1	3	0	0	1	2
	19. pbl + ars + n.e.	0	0	1	17	1	2
Four phase:	20. pbl + qtz + hem + py/mc	4	11	0	0	4	10
	21. pbl + qtz + py/mc + b.m.s.	3	8	0	0	3	7
	22. pbl + carb + py/mc + b.m.s.	2	6	1	17	3	7
	23. pbl + qtz + py/mc + fluor	2	6	0	0	2	5
	24. pbl + qtz + carb + hem	1	3	0	0	1	2
	25. pbl + qtz + carb + py/mc	1	3	0	0	1	2
	26. pbl + qtz + carb + b.m.s.	1	3	0	0	1	2
	27. pbl + qtz + hem + fluor	1	3	0	0	1	2
	28. pbl + carb + hem + py/mc	1	3	0	0	1	2
	29. pbl + carb + hem + sel	1	3	0	0	1	2
	30. pbl + py/mc + b.m.s. + fluor	1	3	0	0	1	2
Five phase:	31. pbl + qtz + carb + hem + py/mc	1	3	0	0	1	2
	32. pbl + qtz + carb + hem + b.m.s.	1	3	0	0	1	2
	33. pbl + qtz + carb + py/mc + b.m.s.	1	3	0	0	1	2
	34. pbl + qtz + hem + py/mc + b.m.s.	1	3	0	0	1	2
	35. pbl + qtz + hem + py/mc + fluor	1	3	0	0	1	2
	36. pbl + carb + hem + py/mc + b.m.s.	1	3	0	0	1	2
	37. pbl + carb + hem + b.m.s. + sel	1	3	0	0	1	2
	38. pbl + carb + py/mc + b.m.s. + ars	0	0	1	17	1	2
Six phase:	39. pbl + qtz + carb + hem + py/mc + b.m.s.	1	3	0	0	1	2

Table 2-4 continued.

Co-deposition pair	35 Deposits without Arsenides		6 Deposits with Arsenides		All 41 Deposits	
	# of deposits	%	# of deposits	%	# of deposits	%
1. pbl + py/mc	25	71	2	33	27	66
2. pbl + qtz	19	54	2	33	21	51
3. pbl + b.m.s.	14	40	1	17	15	36
4. pbl + carb	9	26	1	17	10	24
5. pbl + hem	9	26	1	17	10	24
6. pbl + fluor	3	8	0	0	3	7
7. pbl + ars	0	0	2	33	2	5
8. pbl + n.e.	0	0	2	33	2	5
9. pbl + sel	2	6	0	0	2	5

Abbreviations used:
ars = arsenides
b.m.s. = base metal sulfides
carb = carbonate
fluor = fluorite
hem = hematite
n.e. = native elements
pbl = pitchblende
py/mc = pyrite and/or marcasite
qtz = quartz
sel = selenides

10. In 82% of the deposits containing quartz pitchblende deposition was preceded by quartz.

11. The initial pitchblende stage occurs with or before the deposition of pyrite/marcasite in 68% of the deposits studied.

12. In 94% of the deposits containing a base metal sulfide stage the earliest pitchblende is deposited before or with the sulfide stage.

13. In 83% of the deposits containing cobalt-nickel arsenides the arsenides were deposited after the initial pitchblende stage.

14. In all of the hydrothermal uranium deposits containing both sulfide and arsenide vein stages the major base metal sulfide stage follows the deposition of arsenides.

15. In 67% of the uranium deposits containing carbonate gangue the earliest pitchblende is deposited before the carbonate.

In summary, typical hydrothermal uranium deposits contain pitchblende + pyrite/marcasite. Most deposits also contain quartz, hematite, carbonate, and at least minor quantities of base metal sulfides. The main (first) pitchblende stage is almost always paragenetically early. This pitchblende stage is usually preceded by quartz and hematite deposition and followed by pyrite/marcasite and calcite, an arsenide stage (if present), and finally by a base metal sulfide stage.

A comparison of the paragenetic data for arsenide-bearing and non-arsenide hydrothermal uranium deposits (Tables 2-3 and 2-4) yielded the following generalizations:

1. In a significant proportion (33%) of the arsenide-bearing deposits, the pitchblende stage occupies an intermediate, rather than an early, paragenetic position.

2. Carbonate and barite occur more frequently (83% vs. 46% and 50% vs. 14% respectively) in arsenide-bearing uranium deposits.

3. Initial pitchblende deposition precedes the first appearance of pyrite/marcasite for most (60%) non-arsenide-bearing uranium deposits; pitchblende is deposited with or after pyrite/marcasite, however, for most (67%) of the arsenide-bearing deposits.

4. A base metal sulfide stage is present in all arsenide-bearing deposits, but is absent from 11% of the non-arsenide-bearing deposits.

5. Initial pitchblende deposition precedes the first appearance of carbonate in most (75%) of the non-arsenide uranium deposits containing carbonate, but pitchblende follows carbonate deposition for most (60%) arsenide-bearing deposits containing carbonate.

6. The co-deposition pairs pitchblende + pyrite/marcasite and pitchblende + base metal sulfides occur much more frequently in hydrothermal uranium deposits of simple paragenesis than in arsenide-bearing uranium deposits (71% vs. 33% and 40% vs. 17% respectively).

5. Wall rock alteration

Hematization of wall rocks is characteristic of most but not all important hydrothermal uranium deposits throughout the world (see Table 2-1). In some deposits (e.g. the Fay mine in the Beaverlodge district of Saskatchewan and the Schwartzwalder mine in Colorado) red hematitic wall rock alteration follows uranium veins, but in others no spatial relation between hematization and uranium mineralization is apparent. Although for any single deposit the correlation between the distribution of pitchblende and hematite may be far from perfect, it should be noted that simply the frequent occurrence of primary hematite in hydrothermal uranium deposits serves to distinguish these from the great majority of other hydrothermal ore deposits.

Few published studies of hydrothermal uranium deposits seriously discuss associated wall rock alteration. Boyle (1970)) lists hematization, argillization, carbonatization, silicification, and chloritization as the major wall rock alteration types associated with hydrothermal uranium deposits. Walker (1963), in his review of the wall rock alteration accompanying uranium veins in the United States, adds sericitization and pyritization to Boyle's list. In addition, Walker (1963) states that argillic alteration is the type most frequently reported for hydrothermal uranium deposits. Naumov et al. (1970) report that carbonatization of host rocks during pitchblende deposition is the characteristic alteration type associated with the uranium deposits of the Erzgebirge (complex arsenide-bearing) type. Dawson (1956), referring to the uranium deposits of the Beaverlodge district, Saskatchewan, concludes that amphibole, plagioclase, and biotite were most affected by wall rock alteration, and that hematite, calcite, and albite formed as a result. Petersen (unpublished report, 1973) has noted the rarity of intense wall rock alteration (i.e. advanced argillic, alunitic or pervasive sericitization) associated with hydrothermal uranium deposits, but the deposits of the Marysvale district, Utah, occurring in a shallow volcanic environment, are a notable exception to this generalization.

Boyle (1970) presents the following two generalized wall rock alteration zoning patterns for pitchblende veins:

Type I	Type II
1. Vein	1. Vein
2. Zone of argillic alteration with quartz, sericite, carbonate, and abundant hematite.	2. Zone of albitization with chlorite, sericite, quartz, and abundant hematite.
3. Zone of weak hematitization.	3. Zone of weak hematitization.
4. Unaltered rock.	4. Unaltered rock.

It is important to remember that in any single deposit, one or more (and sometimes all) of these zones may be absent. Walker's (1963) generalized alteration zoning, based primarily on detailed studies of the Caribou mine and deposits of the Central City district, Colorado, the deposits of the Boulder Batholith, Montana, and the deposits of the Marysvale district, Utah, is somewhat different:

1. Vein.
2. Sericite zone with local to widespread silicification.
3. Argillic zone characterized principally by montmorillonite and kaolinite, with hydromica or illite and halloysite or other clay minerals locally important.
4. Chloritic zone with local development of magnetite, ilmenite, pyrite or hematite, calcite, and epidote.
5. Unaltered rock.

6. Summary

Hydrothermal uranium deposits occur most frequently in, or associated with, granitic igneous or regionally metamorphosed rocks. They are characterized by open space filling textures. Hydrothermal uranium deposits have ages ranging from Proterozoic to Tertiary; Archean deposits are unknown. Pitchblende is the only important primary uranium mineral in these deposits. Hydrothermal uranium deposits are divisible into two types on the basis of the complexity of their hypogene paragenesis. Deposits of simple paragenesis contain pitchblende + pyrite/marcasite ± quartz ± hematite ± carbonate ± minor base metal sulfides. Complex deposits contain an important base metal sulfide stage and often possess a cobalt-nickel arsenide stage as well. In most hydrothermal uranium deposits, pitchblende is paragenetically early. Pyrite/marcasite and quartz are the minerals most frequently precipitated together with pitchblende, although pitchblende is commonly deposited alone. Hematite is a characteristic associate of most hydrothermal uranium deposits. The deposition of hypogene vein and/or wall rock alteration hematite usually precedes or accompanies pitchblende deposition.

References for Chapter 2

Badham, J.P.N., Robinson, B.W. and Morton, R.D., 1972, The geology and genesis of the Great Bear Lake silver deposits: 24th Internat. Geol. Congress, Section 4, 541-548.

Boyle, R.W., 1970, Regularities in wall rock alteration phenomena associated with epigenetic deposits: in Pouba, Z. and Štemprok, M., eds., *Problems of Hydrothermal Ore Deposition, 233-260,* Akadémiai Kiadó, Budapest.

Cariou, L., 1964, Régions médiane et Sud du Massif central: in Roubault, M., ed., *Les Minerais Uranifères Français, 3,* pt.1, 9-162, Presses Universitaires de France, Paris.

Carrat, H.G., 1962, Morvan et Autunois: in Roubault, M., ed., *Les Minerais Uranifères Français, 2,* 1-104, Presses Universitaires de France, Paris.

Cuney, M., 1974, Le gisement uranifère des Bois-Noirs-Limouzat (Massif Central-France) — Relations entre minéraux et fluides: unpub. doctoral thesis, Nancy, 174 pp.

Dawson, K.R., 1956, Petrology and red coloration of wall-rocks, radioactive deposits, Goldfields region, Saskatchewan: Canada Geol. Survey Bull. 33, 46 pp.

Drake, A.A., Jr., 1957, Geology of the Wood and East Calhoun mines, Central City district, Gilpin County, Colorado: U.S. Geol. Survey Bull. 1032-C, 129-170.

Everhart, D.L. and Wright, R.J., 1953, The geologic character of typical pitchblende veins: Econ. Geology, *48,* 77-96.

Geffroy, J. and Sarcia, J.A., 1955, Contribution à l'étude des pechblendes françaises: Rapport C.E.A. no. 380, 157 pp.

Geffroy, J. and Sarcia, J.A., 1960, Les minerais noirs: in Roubault, M. ed., *Les Minerais Uranifères Français, 1,* 1-68, Presses Universitaires de France, Paris.

Germain, C., Kervella, M. and Le Bail, F., 1964, Bretagne: in Roubault, M. ed., *Les Minerais Uranifères Français, 3,* pt. 1, 209-275, Presses Universitaires de France, Paris.

Gornitz, V. and Kerr, P.F., 1970, Uranium mineralization and alteration, Orphan mine, Grand Canyon, Arizona: Econ. Geology, *65,* 751-768.

Gross, E.B., 1965, A unique occurrence of uranium minerals, Marshall Pass, Saguache County, Colorado: Am. Mineralogist, *50,* 909-923.

Naumov, G.B., Acheyev, B.N. and Yermolayev, N.P., 1970, Movement of hydrothermal solutions: Internat. Geology Rev., *12,* 610-618.

Naumov, G.B., Motorina, Z.M. and Naumov, V.B., 1971, Conditions of formation in veins of the Pb-Co-Ni-Ag-U type: Geochem. International, *6,* 590-598.

Petersen, U., 1973, Hydrothermal uranium deposits: unpub. report prepared for Exxon Corporation.

Robinson, B.W. and Badham, J.P.N., 1974, Stable isotope geochemistry and the origin of the Great Bear Lake silver deposits, N.W.T., Canada: Canadian Jour. Earth Sci., *11,* 696-711.

Robinson, B.W. and Ohmoto, H., 1973, Mineralogy, fluid inclusions, and stable isotopes of the Echo Bay U-Ni-Ag-Cu deposits, Northwest Territories, Canada: Econ. Geology, *68,* 635-656.

Robinson, S.C., 1955, Mineralogy of uranium deposits, Goldfields, Saskatchewan: Canada Geol. Survey Bull. 31, 128 pp.

Ruzicka, V., 1971, Geological comparison between East European and Canadian uranium deposits: Canada Geol. Survey Paper 70-48, 196 pp.

Sassano, G.P., Fritz, P. and Morton, R.D., 1972, Paragenesis and isotopic composition of some gangue minerals from the uranium deposits of Eldorado, Saskatchewan: Canadian Jour. Earth Sci., *9,* 141-157.

Smith, E.E.N., 1952, Structure, wall-rock alteration and ore deposits at Martin Lake: unpub. doctoral thesis, Harvard, 125 pp.

Smith, E.E.N., 1974, Review of current concepts regarding vein deposits of uranium: in *Formation of Uranium Ore Deposits,* 515-529, Internat. Atomic Energy Agency, Vienna.

Walker, G.W., 1963, Host rocks and their alterations as related to uranium-bearing veins in the conterminous United States: U.S. Geol. Survey Prof. Paper 455-C, 37-53.

Walker, G.W. and Adams, J.W., 1963, Mineralogy, internal structural and textural characteristics, and paragenesis of uranium-bearing veins in the conterminous United States: U.S. Geol. Survey Prof. Paper 455-D, 55-90.

Walker, G.W. and Osterwald, F.W., 1963, Introduction to the geology of uranium-bearing veins in the conterminous United States, including sections on geographic distribution and classification of veins: U.S. Geol. Survey Prof. Paper 455-A, 1-28.

Chapter 3
REVIEW OF FLUID INCLUSION STUDIES OF HYDROTHERMAL URANIUM DEPOSITS

1. Introduction

A substantial amount of fluid inclusion data pertaining to hydrothermal uranium deposits is contained in the literature. Unfortunately much of the data are of only limited value for determining pressure and temperature during pitchblende deposition because homogenization and decrepitation temperatures are frequently reported without accompanying estimates of fluid composition, and also because many of the fluid inclusions are hosted by minerals which were not deposited simultaneously with pitchblende. However, several particularly useful studies have been published; they are discussed below. The results of these and other relevant fluid inclusion studies are summarized in Tables 3-1 and 3-2.

2. Fluid inclusion studies related to the deposition of pitchblende

To date, the most complete fluid inclusion studies of hydrothermal uranium deposits have been made by B. Poty and his co-workers at the Centre de Recherches Pétrographiques et Géochimiques in Nancy, France. The results of these studies are presented in Poty et al. (1974), which summarizes and updates the work reported by Cuney (1974) and Leroy and Poty (1969) on hydrothermal uranium deposits of the Massif Central in France.

Fluid inclusion studies of the Bois Noirs-Limouzat deposit in the Forez region of the Massif Central have characterized, at least in preliminary fashion, the hydrothermal fluids present during each of the 6 paragenetic stages of vein deposition (Figure 3-1) carefully described by Cuney (1974) and Arnold and Cuney (1974). It has been clearly demonstrated by Cuney (1974) that the composition of the hydrothermal fluids varied dramatically during the formation of the Limouzat vein at Bois Noirs (Figure 3-2b). The wide range of homogenization temperatures for the several stages suggests that the temperature of vein formation may also have varied significantly (Figure 3-2a). It should be mentioned, however, that the very broad range of homogenization temperatures for Stage II fluid inclusions is probably explained by the boiling of a 150-200°C fluid.

Low temperature, dilute aqueous fluids of variable but elevated CO_2 content were present during the period of pitchblende deposition (Stage I) in the Limouzat vein. Samples of these fluids are contained in one and two phase primary fluid inclusions in Stage I quartz which was deposited simultaneously with early pitchblende. Some of these fluid inclusions contain fibrous masses of birefringent crystals (thought to be the calcium analogue of dawsonite) which commonly occupy as much as 50 volume % of the inclusion cavities. It is likely that these crystal aggregates were accidentally incorporated in the inclusions and were not precipitated from the inclusion fluids subsequent to their

Table 3-1. Fluid inclusion data pertaining to the deposition of pitchblende in hydrothermal uranium deposits

Deposit or Region	Filling T (°C)	Decrepitation T (°C)	Formation T (°C)	Formation P (bars)	CO_2 (Mole %)	Salinity (eq. wt. % NaCl)	Reference
Great Bear Lake, Canada	~150	-	-	-	-	-	Badham et al. (1972).
Erzgebirge	80-143	-	-	-	0-5.1	-	Naumov & Mironova (1969); Naumov et al. (1971).
"	35-165	50-180; pitchblende	-	-	-	-	Tugarinov and Naumov (1969).
Western Erzgebirge	-	-	80-180	-	-	-	**Harlass & Schüetzel (1965).**
Forez, France	25-150	-	-	-	1.92*	1.7*	Cuney (1974); Poty et al. (1974).
Limousin, France	≥300	>380; quartz	340-350	700-800	3-20**	5-15	Leroy & Poty (1969); Poty et al. (1974).
Oberpfalz, Germany	-	-	-	-	0.3-1.9	-	Kranz (1968).
U.S.S.R., Unidentified deposit(s)	112-145	130-150; calcite	-	-	-	-	Barsukov et al. (1971)
" "	-	-	185-220	-	-	-	Kotov et al. (1968, 1970).
" "	-	-	220-230	-	-	-	Oparysheva et al. (1974).
" "	-	-	150-200	-	**	-	Rogova et al. (1971).
" "	136-168	-	-	-	-	-	Tugarinov & Naumov (1969).
" "	170-190	-	-	-	-	-	"
" "	124-168	-	-	-	-	-	"
" "	102-135	100-145; fluorite, pitchblende, molybdenite	-	480-500	-	-	"
" "	90-146	120-160; galena, pitchblende	-	480-500	**	-	"
" "	78-96	100 niccolite, pitchblende	-	-	-	-	"

*Average value for pitchblende stage.
**Liquid CO_2-bearing fluid inclusions seen in samples studied, but not necessarily in pitchblende stage samples.

Table 3-2. Additional fluid inclusion data related to hydrothermal uranium deposits

Deposit or Region	Stage	Filling T (°C)	Decrepitation T (°C)	Formation T (°C)	Formation P (bars)	CO_2 mole %	Salinity. (equiv. wt. % NaCl)	Reference
Great Bear Lake district	-	-	≤400	-	-	-	-	Campbell (1955)
" "	Post-U	110-160	-	140-230	300-800	-	~30	Robinson & Ohmoto (1973)
" "	Post-U	-	-	180-510	-	-	30-35	Shegelski & Scott (1975)
Beaverlodge district	-	-	256-400; quartz, calcite	-	-	-	-	Robinson (1955)
" "	Pre-U	145-195*	-	75-470	-	-	-	Sassano et al. (1972)
	Post-U	60-410	-		<300	-	≤26-28	"
Rabbit Lake deposit, Saskatchewan	Post-U	-	<100; quartz, calcite	-	-	-	-	Knipping (1974)
" "	Pre-U	245	-	-	-	-	-	Little (1974)
" "	Post-U	135-160	-	≥180-225	-	-	-	"
" "	Post-U	92-182	-	>130-170	>300-900	**	~28-30	Pagel (1976)
Orphan mine, Arizona	-	45-124	-	-	-	-	-	Gornitz and Kerr (1970)

* Data for secondary fluid inclusions.
** CO_2 and hydrocarbons present.

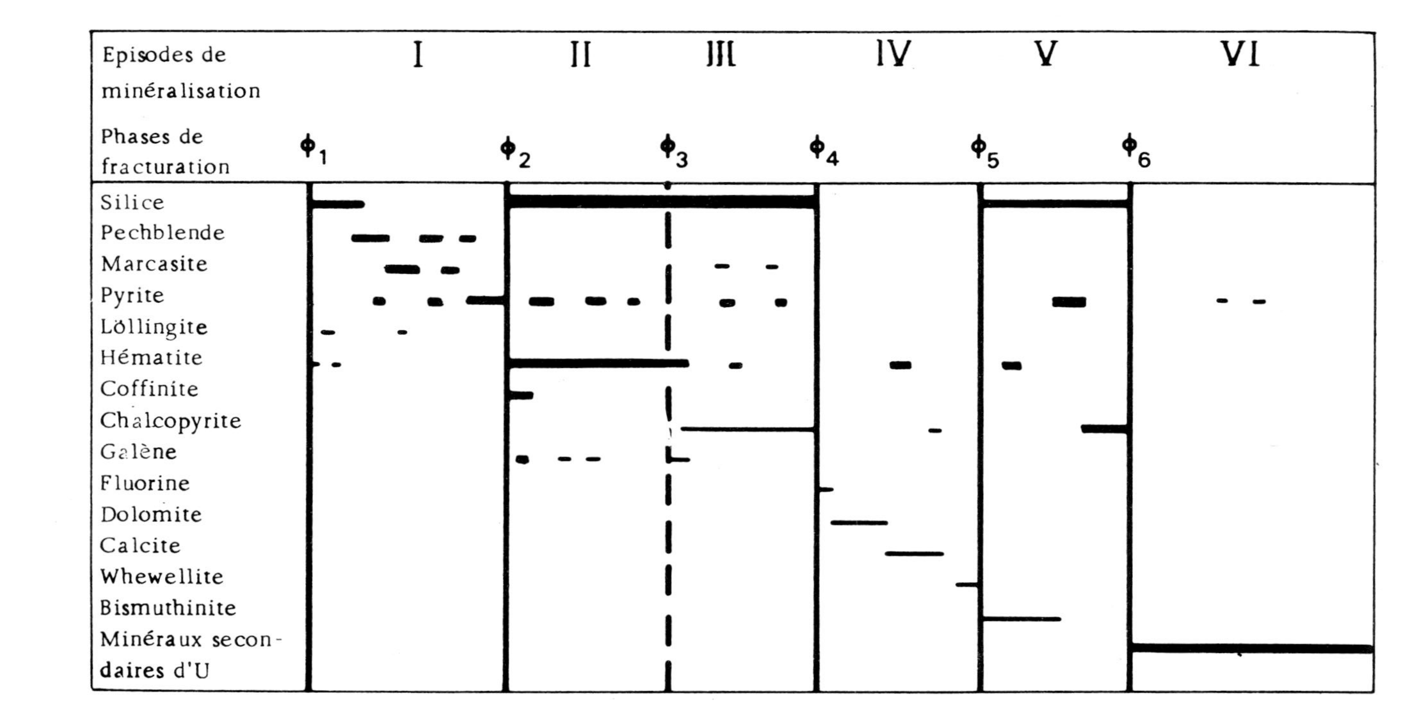

Fig. 3-1. Paragenesis of the Bois-Noirs-Limouzat deposit, Forez region, France (from Poty et al., 1974).

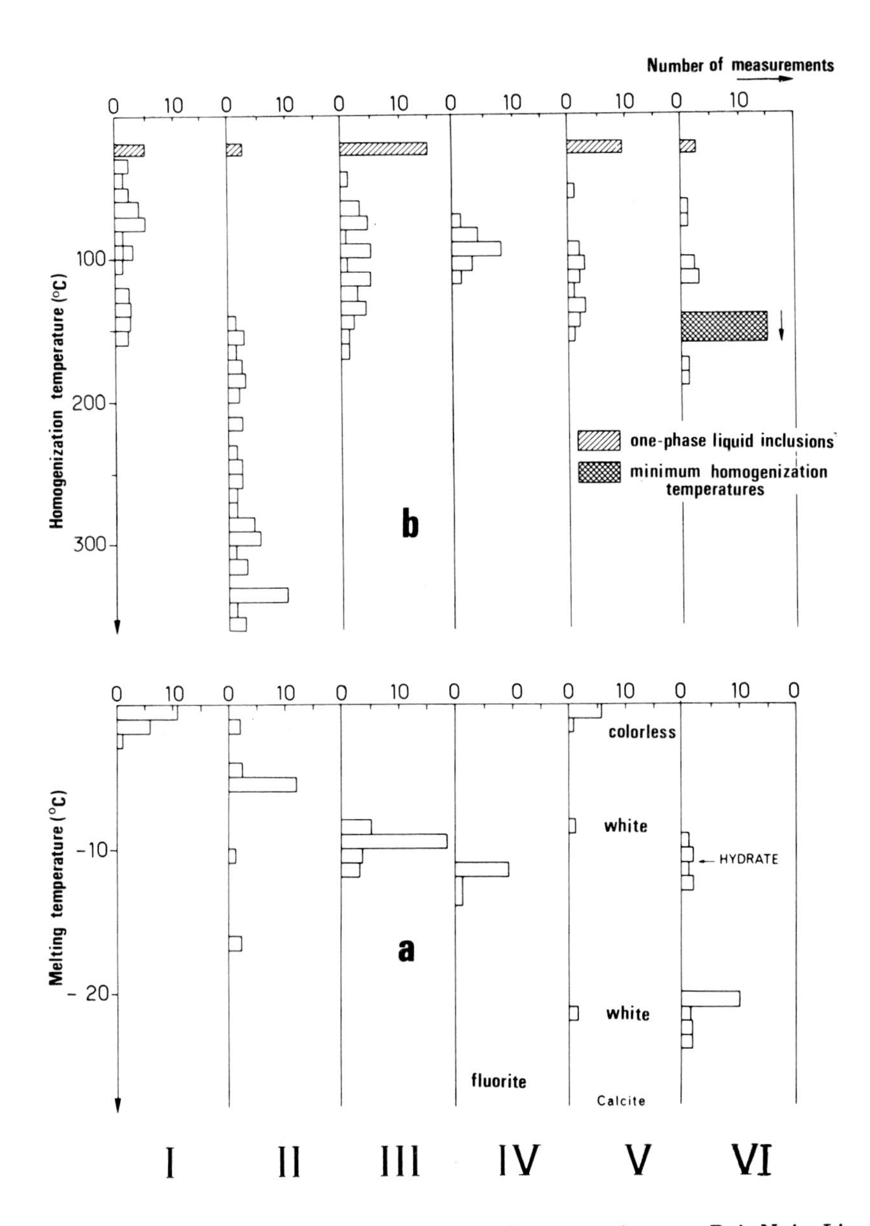

Fig. 3-2. Fluid inclusion data as a function of paragenetic stage, Bois-Noirs-Limouzat deposit, Forez region, France (modified from Poty et al., 1974).

entrapment. Primary Stage I fluid inclusions are characterized by very low filling temperatures (50 to 150° C) and low salinities (1-2 equivalent weight % NaCl). Crushing stage work revealed a great variation in the content of non-condensible volatiles (chiefly CO_2) in the inclusion fluids. An average analysis of the volatiles contained in Stage I quartz showed the presence of 2.6 mole % non-condensible gases, of which 1.9% was CO_2 and 0.7% hydrocarbons; the presence of a liquid CO_2 phase was never observed in fluid inclusions of the pitchblende stage. Water accounted for the remaining 97.4 mole % of the volatile content of average Stage I fluid. The low homogenization temperatures and the presence of CO_2 and Ca-dawsonite (?) which characterize Stage I fluid inclusions at Bois Noirs-Limouzat have also been observed for inclusions in ore samples from Jáchymov, Rabbit Lake, and various uranium prospects throughout the Forez region. (Cuney, 1976).

During post-pitchblende vein formation (Stages II through VI) at Limouzat, the P-T-X conditions of the hydrothermal fluids were quite variable (Poty et al., 1974). Inclusion filling temperatures range from 25 to 350° C, and inclusion fluids contain between 1.7-23 equivalent weight % NaCl and 0 to 7.7 mole % CO_2. It is particularly interesting that Cuney (1974) noted the existence of hypogene leaching of pitchblende during Stage V; he ascribed this to the presence of elevated CO_2 contents in late Stage IV and Stage V fluids. Cuney (1974) suggests that boiling of the fluids may have taken place during Stage II. This would explain the wide range of filling temperatures and fluid salinities recorded for that period of vein deposition.

Fluid inclusion studies of the Margnac and Fanay uranium deposits in the Limousin region of the Massif Central are also discussed by Poty et al. (1974). Quantitative data are presently available only for the episyenite uranium ores; the pitchblende vein occurrences are presently being investigated (see Leroy, 1976). Leroy and Poty (1969) studied primary and secondary fluid inclusions in relict magmatic quartz and hydrothermal chalcedony spatially associated with pitchblende in episyenite. Although it can not be conclusively proven that these fluid inclusions formed during the deposition of pitchblende, Poty et al. (1974) have demonstrated that a clear relation exists between the grade of uranium ore in episyenite and the CO_2 concentration of associated inclusion fluids (Figure 3-3). Such a relation suggests that the fluid inclusions studied are co-genetic with pitchblende. If so, the deposition of pitchblende in the Limousin region took place at a temperature of 340-350° C and a pressure of 700-800 bars from low to moderate salinity fluids rich in CO_2. The high formation temperature suggested by Poty et al. (1974) for the Margnac and Fanay pitchblendes is unique among the results of the several fluid inclusion studies of hydrothermal uranium deposits (see Table 3-1). The wide range of filling temperatures and CO_2/H_2O ratios exhibited by the Limousin fluid inclusions is cited as evidence of fluid boiling during uranium mineralization (Leroy and Poty, 1969). No other study of hydrothermal uranium deposits has reported the existence of boiling during the stage of pitchblende deposition.

The fluid inclusion data of Naumov and Mironova (1969) indicate a progressive decrease of CO_2 (from 4.0 to 0.2 mole %) in the hydrothermal fluids present during deposition of the quartz-pitchblende-calcite stage in hydrothermal uranium deposits of

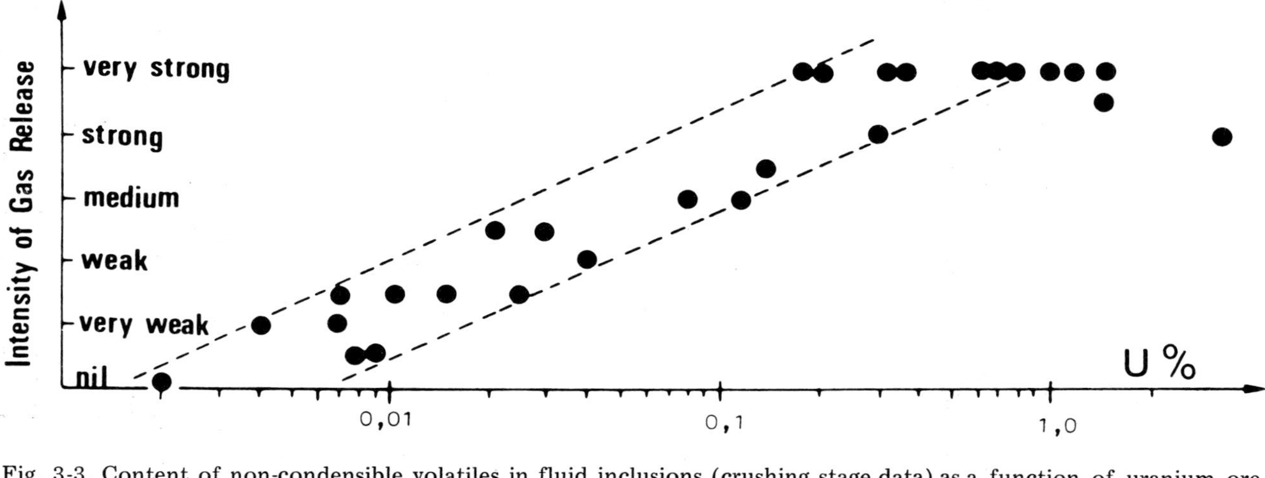

Fig. 3-3. Content of non-condensible volatiles in fluid inclusions (crushing stage data) as a function of uranium ore grade, Margnac deposit, Limousin region, France (modified from Poty et al., 1974).

the Erzgebirge region.* The investigations of Naumov and Mironova (1969) were extended by Naumov et al. (1971) to include other paragenetic stages of the Erzgebirge deposits; filling temperatures were also determined. The data of Naumov and Mironova (1969) and Naumov et al. (1971) are summarized in Table 3-3. Note the uniformly low filling temperatures (80-156°C) for major vein Stages 3-5.

Table 3-3. Fluid inclusion data of Naumov and Mironova (1969) and Naumov et al. (1971) for hydrothermal uranium deposits of the Erzebirge.

Stage	Inclusion Host Mineral	Filling T (°C)	mole % CO_2*
1. Early (pre-vein) quartz-sulfide	quartz	201-312	-
2. Comb quartz of vein margins	quartz	122-175	-
3. Quartz-pitchblende -calcite	quartz	-	4.0±1.1
	pitchblende	-	0.7±0.2
	calcite	80-143	0.2±0.2
4. Dolomite-selenide	dolomite	110-145	0.5±0.4
5. Arsenide	ankerite	114-156	1.1±0.4
	siderite	92-137	1.6±0.7
6. Late quartz-sulfide -calcite	calcite	<51	trace

* Mean value ± standard deviation (90% confidence).

Tugarinov and Naumov (1969) present the results of an extensive study of the fluid inclusion geothermometry of several unidentified vein-type pitchblende deposits in the U.S.S.R. and in the Erzgebirge region of central Europe. Filling temperatures for gangue minerals from the pitchblende stages of the various deposits range from 35 to 190°C, and decrepitation temperatures range from 50 to 180°C for several vein minerals including pitchblende itself. The data of Tugarinov and Naumov (1969) for the individual pitchblende deposits are presented in Table 3-1. Tugarinov and Naumov (1974) summarize the data of previous Russian fluid inclusion studies concerning hydrothermal uranium deposits, but the data presented in this summary differ little from those reported by Tugarinov and Naumov (1969).

* The same effect has recently been qualitatively documented by Leroy (1976) for the pitchblende veins of the Fanay deposit in the Limousin region of France.

3. Other fluid inclusion studies of hydrothermal uranium deposits

Other fluid inclusion studies of hydrothermal uranium deposits have been reported by Campbell (1955), Robinson (1955), Gornitz and and Kerr (1970), Sassano et al. (1972), Robinson and Ohmoto (1973), Knipping (1974), Little (1974), Shegelski and Scott (1975), and Pagel (1976). These studies, however, do not explicitly present fluid inclusion data for the stage of pitchblende deposition. Nevertheless, the results of these investigations are summarized in Table 3-2 because they are of interest with regard to the geochemical evolution of the mineralization systems.

4. Discussion and conclusions

With the notable exception of the results of Leroy and Poty (1969), filling temperatures for the pitchblende stages of all hydrothermal uranium deposits studied are $\leqslant$190°C; most (7 of 12) of these deposits have filling temperatures $\leqslant$150°C (Table 3-1). Although filling temperatures, in general, are lower than formation temperatures, the epithermal nature of many hydrothermal uranium deposits and the low ($\leqslant$1 kb) formation pressures given in Tables 3-1 and 3-2 suggest that the formation temperatures were not markedly higher than the filling temperatures. The estimated formation temperatures given by various workers for the pitchblende stage of several hydrothermal uranium deposits (Table 3-1) support this conclusion. The data of Leroy and Poty (1969) are an exception. The reported filling temperatures of 300°C and above and formation temperatures of 340-350° C for the pitchblende stage exceed by at least 100°C the corresponding data for all other hydrothermal uranium deposits. In view of this disparity, it is interesting to note that only Leroy and Poty (1969) present evidence for fluid boiling during pitchblende deposition. In addition, their study is also the only one which dealt with uranium mineralization in an episyenite.

Data for the composition of fluid inclusions associated with pitchblende deposition are limited; nevertheless some generalizations regarding the nature of the fluids in these inclusions may be ventured. The data given in Table 3-1 and the absence of daughter crystals in most fluid inclusions of the pitchblende stage suggest that the fluids depositing pitchblende have low to moderate salinities. The CO_2 content of pitchblende stage fluids is generally $\geqslant$ 1 mole %. Fluid inclusions containing liquid CO_2 have been reported in vein material from several hydrothermal uranium deposits, and CO_2 hydrate has been observed during freezing stage runs in fluid inclusions from additional deposits. Naumov and Mironova (1969) describe a pronounced decrease in the CO_2 content of fluids during the pitchblende stage of vein formation in the Erzgebirge (see Table 3-3). The extremely CO_2-rich (up to 20 mole %) fluid inclusions associated with pitchblende deposition at Limousin are probably products of fluid boiling (Leroy and Poty, 1969).

In summary, fluid inclusion studies of hydrothermal uranium deposits suggest that pitchblende is usually deposited at low to intermediate temperatures and pressures from CO_2-bearing, aqueous fluids of low salinity. The variable temperature and composition data for the fluids of multi-stage uraniferous veins (Cuney, 1974; Robinson and

Ohmoto, 1973) indicate that the conditions accompanying pitchblende deposition can be inferred reliably only from the properties of contemporaneous fluid inclusions.

References for Chapter 3

Arnold, M. and Cuney, M., 1974, Une succession anormale de minéraux et ses conséquences sur l'exemple de la minéralisation uranifère des Bois Noirs-Limouzat (Forez-Massif Central français): Acad. Sci. Comptes Rendus, Ser. D, *279*, 535-538.

Badham, J.P.N., Robinson, B.W. and Morton, R.D., 1972, The geology and genesis of the Great Bear Lake silver deposits: 24th Internat. Geol. Congress, Section 4, 541-548.

Barsukov, V.L., Sushchevskaya, T.M. and Malyshev, B.I., 1971, Composition of solutions forming pitchblende in a uranium-molybdenum deposit (abst.): Fluid Inclusion Research — Proceedings of C.O.F.F.I., *4,* 10, privately published, Washington, D.C.

Campbell, D.D., 1955, Geology of the pitchblende deposits of Port Radium, Great Bear Lake, N.W.T.: unpub. doctoral thesis, Cal. Inst. of Tech.

Cuney, M., 1974, Le gisement uranifère des Bois-Noirs-Limouzat (Massif Central — France) — Relations entre minéraux et fluides: unpub. doctoral thesis, Nancy, 174 pp.

Cuney, M., 1976, Le gisement d'uranium des Bois-Noirs: Rapport Annuel 1975, Centre de Recherches Pétrographiques et Géochimiques, 49-50.

Gornitz, V. and Kerr, P.F., 1970, Uranium mineralization and alteration, Orphan Mine, Grand Canyon, Arizona: Econ. Geology, *65,* 751-768.

Harlass, E. and Schützel, H., 1965, Zur paragenetischen Stellung der Uranpechblende in den hydrothermalen Lagerstätten des westlichen Erzgebirges: Zeitschr. Angew. Geologie, *11*, 569-581.

Knipping, H.D., 1974, The concepts of supergene versus hypogene emplacement of uranium at Rabbit Lake, Saskatchewan, Canada: in *Formation of Uranium Ore Deposits*, 531-549, Internat. Atomic Energy Agency, Vienna.

Kotov, Ye. I., Timofeev, A.V. and Khoteev, A.D., 1968, Formation temperatures of minerals of uranium hydrothermal deposits (abst.): Fluid Inclusion Research — Proceedings of C.O.F.F.I., *1*, 44, privately published, Washington, D.C.

Kotov, Ye. I. et al., 1970, Formation temperatures of some hydrothermal uranium deposits: Fluid Inclusion Research — Proceedings of C.O.F.F.I., *3*, 38, privately published, Washington, D.C.

Kranz, R.L., 1968, Participation of organic compounds in the transport of ore metals in hydrothermal solutions: Inst. Mining and Metallurgy Trans., Section B, *77*, B26-B36.

Leroy, J., 1976, Les structures filoniennes mineralisées en pechblende de Fanay: Rapport Annuel 1975, Centre de Recherches Pétrographiques et Géochimiques, 47-49.

Leroy, J. and Poty, B., 1969, Recherches préliminaires sur les fluides associés à la genèse des minéralisations en uranium du Limousin (France): Mineralium Deposita, *4*, 395-400.

Little, H.W., 1974, Uranium in Canada: Canada Geol. Survey Paper 74-1A, 137-139.

Naumov, G.B. and Mironova, O.F., 1969, Das Verhalten der Kohlensäure in hydrothermalen Lösungen bei der Bildung der Quarz-Nasturan-Kalzit-Gänge des Erzgebirges: Zeitschr. Angew. Geologie *15*, 240-241.

Naumov, G.B., Motorina, Z.M. and Naumov, V.B., 1971, Conditions of formation of carbonates in veins of the Pb-Co-Ni-Ag-U type: Geochem. International, *6*, 590-598.

Oparysheva, L.G., Shmariovich, Ye. M., Larkin, E.D. and Shchetochkin, V., 1974, Characteristics of uranium mineralization in sedimentary blanket and basement granites: Internat. Geology Rev., *16*, 600-609.

Pagel, M., 1976, Conditions de dépôt des quartz et dolomites automorphes du gisement uranifère de Rabbit Lake (Canada): Réunion Annuelle des Sciences de la Terre, Paris (in press).

Poty, B.P., Leroy, J. and Cuney, M., 1974, Les inclusions fluides dans les minerais des gisements d'uranium intragranitiques du Limousin et du Forez (Massif Central, France): in *Formation of Uranium Ore Deposits*, 569-582, Internat. Atomic Energy Agency, Vienna.

Robinson, B.W. and Ohmoto, H., 1973, Mineralogy, fluid inclusions, and stable isotopes of the Echo Bay U-Ni-Ag-Cu deposits, Northwest Territories, Canada: Econ. Geology, *68,* 635-656.

Robinson, S.C., 1955, Mineralogy of uranium deposits, Goldfields, Saskatchewan: Canada Geol. Survey Bull. 31, 128 pp.

Rogova, V.P., Nikitin, A.A.and Naumov, G.B., 1971, Mineralogic-geochemical conditions of the localization of U-Mo deposits in volcanogenic sedimentary formations (abst.): Fluid Inclusion Research — Proceedings of C.O.F.F.I., *4*, 68-69, privately published, Washington, D.C.

Sassano, G.P., Fritz, P. and Morton, R.D., 1972, Paragenesis and isotopic composition of some gangue minerals from the uranium deposits of Eldorado, Saskatchewan: Canadian Jour. Earth Sci., *9*, 141-157.

Shegelski, R.J. and Scott, S.D., 1975, Geology and mineralogy of the Ag-U-arsenide veins of the Camsell River district, Great Bear Lake, N.W.T. (abst.): Geol. Soc. America Abs. with Programs, *7*, 857-858.

Tugarinov, A.I. and Naumov, G.B., 1974, Die Migrations- und Absatzverhältnisse des Urans bei der endogenen Erzbildung: Zeitschr. Angew. Geologie, *20,* 410-413.

Tugarinov, A.I. and Naumov, V.B., 1969, Thermobaric conditions of formation of hydrothermal uranium deposits: Geochem. International, *6*, 89-103.

Chapter 4
THE CHEMISTRY OF URANIUM TRANSPORT IN HYDROTHERMAL FLUIDS

1. Introduction

The chemistry of hydrothermal uranium deposits is similar to the chemistry of zinc and lead deposits. Within each class of deposits a single ore mineral is overwhelmingly important. Uranium deposits are, however, complicated by the existence of two geologically important valence states of uranium and by the variety of the uranium complexes which are probably present in ore-forming hydrothermal fluids.

Much of our present thinking regarding the geochemistry and origin of uranium deposits is derived from the work of R.M. Garrels and his colleagues during the fifties and early sixties (e.g. Garrels, 1955; Garrels and Christ, 1959; Garrels and Larsen, 1959; Garrels and Pommer, 1959; Garrels and Richter, 1955; Hostetler and Garrels, 1962; McKelvey, Everhart, and Garrels, 1955). Although their interest focused largely on the problems posed by the geology of the uranium deposits of the Colorado Plateau, their approach is directly applicable to the study of hydrothermal deposits. Nearly all of the thermochemical data used to construct their well known Eh-pH diagrams were limited to 25°C and a pressure of 1 atm. Experiments since the fifties, particularly those of several groups in the U.S.S.R. and more recently those of the group at the Centre de Recherches Pétrographiques et Géochimiques in Nancy, France, have begun to supply some of the data required for an understanding of the origin of uranium deposits formed at temperatures above 100°C. A great deal of work, however, remains to be done, and the treatment given in this chapter is somewhat less than satisfactory because only a small fraction of the necessary chemical data are currently available.

2. Uraninite and the system U-O

Uraninite is the only important hypogene uranium mineral present in hydrothermal deposits. In these deposits the particle size of uraninite is usually less than 10^{-2} mm and may be smaller than 10^{-6} mm. The varietal name pitchblende is applied to such fine-grained uraninite. Pitchblende commonly exhibits a botryoidal habit, and its composition is rather variable. (see for instance Frondel, 1958). Natural pitchblendes range in composition from $UO_{2.0}$ to $UO_{2.6}$. With increasing U^{+6} content the unit cell size of pitchblende decreases and the density increases; this suggests that charge balance is maintained in the structure by the presence of additional 0^{-2} ions rather than by a reduction in the number of uranium ions. Hydrothermal uraninite generally contains only small quantities of the rare earths and thorium.

Isometric uraninite is the only anhydrous uranium oxide which has been observed in nature. The presence of only a single phase spanning the entire compositional range from $UO_{2.0}$ to $UO_{2.6}$ is rather surprising in view of the complexity of the experimentally determined U-O phase diagram for elevated temperatures. In the laboratory, solid solution between UO_2 and UO_3 is usually found to extend no further than $UO_{2.2}$ or $UO_{2.3}$. When $UO_{2.0}$ is oxidized in air at temperatures between 25°C and 250°C compositional gaps have been observed between $UO_{2.3}$ and U_3O_8 ($UO_{2.67}$) and between $UO_{2.67}$ and $UO_{3.0}$. (See for instance Figure 4-1 from Pério, 1953). Thermochemical data for uranium oxides with a variety of O/U ratios have been reported during the past decade (see for instance Leitnaker and Godfrey, 1966; Fitzgibbon et al., 1967; Affortit, 1969; Fredrickson and Chasanov, 1970; Gronvold et al., 1970; Ackermann and Chang, 1973), and these in part supersede the compilation of Rand and Kubaschewski (1963).

The difference between the phase relations observed in laboratory studies of the U-O system and those found in naturally occurring pitchblendes is somewhat puzzling. The wide range of U/O ratios in natural pitchblendes could be explained by non-equilibrium crystallization during ore deposition, but the persistence of UO_{2+x} solid solutions in the field suggests that they may be closer to equilibrium than the low-temperature uranium oxides synthesized in the laboratory.

3. Solid phases in the system UO_3-H_2O-CO_2

Mineral phases in this system seem to be restricted to zones of secondary alteration. They are of interest in the study of hydrothermal uranium deposits only because they define upper limits of f_{CO_2} and f_{O_2} for the hydrothermal solutions which deposited pitchblende. The mineralogy of the UO_3 hydrates is still somewhat obscure. Becquerelite seems to have a composition in the range $UO_3 \cdot 1.5$-$1.6\ H_2O$ and schoepite in the range $UO_3 \cdot 2.0$-$2.5\ H_2O$. A rather large number of UO_3 hydrates have been produced in the laboratory, but their dehydration sequence is only vaguely known.

The upper limit of thermal stability of $UO_3 \cdot 2H_2O$ in contact with liquid water is about 60°C. Under a constant H_2O pressure of 15mm, $UO_3 \cdot 2H_2O$ gradually loses ½ H_2O between 30° and 100°C. Another ½ H_2O is lost at 100°C, a third ½ H_2O gradually between 100°C and 300°C, and the remaining ½ H_2O at 300°C (Frondel, 1958, p. 77). $UO_3 \cdot H_2O$ seems to be stable in the presence of liquid water at least up to 200°C (Sergeyeva et al., 1972).

The stable uranium phase in the presence of liquid water at 25°C is probably schoepite ($UO_3 \cdot 2.0 - 2.5\ H_2O$), although the free energy of dehydration to a lower hydrate, probably $UO_3 \cdot H_2O$, is very small. The only carbonate of UO_3 is the mineral rutherfordine (UO_2CO_3). All of the other uranyl carbonates contain at least one other essential cation.

There has been some debate regarding the CO_2 pressure at which rutherfordine is in equilibrium with schoepite and liquid water at 25°C. The work of Sergeyeva et al. (1972) suggests that at 25°C the equilibrium CO_2 pressure is $10^{-2.2}$ atm for the assemblage UO_2CO_3, $UO_3 \cdot H_2O$ and water. At 100, 150 and 200°C the CO_2 pressure for equilibrium among these phases was found to be 1.6 ± 0.5, 4.5 ± 0.5 and 30 ± 1 atm respectively. Since $UO_3 \cdot H_2O$ is nearly in equilibrium with $UO_3 \cdot 2H_2O$ at 25°C, the value of $10^{-2.2}$ atm has been taken as the best estimate of the CO_2 pressure for the rutherfordine-schoepite-water equilibrium at 25°C.

The solubility data of Sergeyeva et al. (1972) yield a free energy of formation of

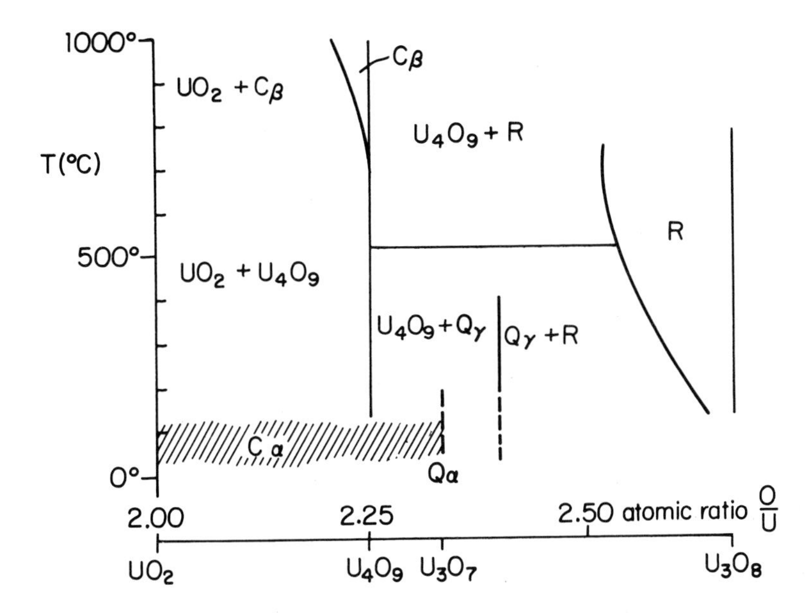

Fig. 4-1. Probable phase relations for U-O system between UO_2 and U_3O_8; shaded area represents compositional range of cubic α phase; C = cubic, R = orthorhombic, and Q = tetragonal (redrawn from Perio, 1953).

-382.4 kcal/mol for rutherfordine if the Hostetler and Garrels (1962) value of -236.4 kcal/mol for the free energy of formation of the uranyl ion (UO_2^{2+}) is used. This result is in reasonable agreement with the -385.0 kcal/mol value reported by Hostetler and Garrels (1962). The free energy of $UO_3 \cdot H_2O$ calculated from the data of Sergeyeva et al. (1972) is then -398.5 kcal/mol, which is in excellent agreement with the value of -398.8 kcal/mol obtained by Hostetler and Garrels (1962). The log f_{O_2} - log f_{CO_2} diagram of Figure 4-2 summarizes the available data for the relevant mineral phases in the system U-O-C-H at 25°C. The field of uraninite is bounded by that of graphite on the low-f_{O_2} side and by that of schoepite and rutherfordine on the high-f_{O_2} side. Figure 4-2 is somewhat inaccurate, because insufficient thermochemical data are available to plot the stability fields of the uranium oxides between $UO_{2.0}$ and $UO_{2.6}$. In future revisions the UO_{2+x}-schoepite and the UO_{2+x}-rutherfordine boundaries will probably be shifted somewhat toward higher values of f_{O_2}.

4. Uranium concentrations in aqueous solutions saturated with respect to minerals in the system U-O-H_2O

Currently available data for the solubility of uranium minerals are surprisingly poor, even at 25°C. Figure 4-3 shows the data of Gayer and Leider (1955) and Miller (1958) for the solubility of $UO_3 \cdot H_2O$ in perchloric acid at 25°C and at pH values between 4 and 5. Although there is a great deal of scatter in the data and the curve defined by the equation

$$\Sigma m_U = 60\ a_{H^+} + 3.6 \times 10^5 a_{H^+}^2$$

is not very precise, the data suggest that at higher pH values the dominant uranium complex is monovalent whereas at lower pH it is divalent. It seems likely that the monovalent complex is $UO_2(OH)^+$ and the divalent complex UO_2^{+2}.

The solubility reported by Gayer and Leider (1955) for $UO_3 \cdot H_2O$ in water is 3.95 x 10^{-5} mol U/1000 gm H_2O (10 ppm). Because this value is well in excess of the extrapolation of the curve in Figure 4-3 to a pH of 7.0, it is likely that the presence of a neutral complex, perhaps $UO_2(OH)_2^\circ$, accounts for the bulk of the $UO_3 \cdot H_2O$ solubility at and near neutral pH values. The solubility of $UO_3 \cdot H_2O$ increases very slowly with increasing pH in alkaline media. Gayer and Leider (1955) report that the addition of 4.5 x 10^{-3} mol NaOH/1000gm H_2O only increases the solubility of $UO_3 \cdot H_2O$ to 4.75 x 10^{-5} mol U/1000 gm H_2O (11 ppm). Ricci and Loprest (1955) reported similar solubilities of $UO_3 \cdot H_2O$ (3-7 x 10^{-5} mol U/l) for solutions in the system Na_2O-UO_3-H_2O at 50 and 75°C. It seems likely, therefore, that the solubility of $UO_3 \cdot H_2O$ at all geologically reasonable pH values greater than 7 is nearly equal to its solubility in neutral solutions.

The solubility of uraninite at 25°C has not yet been measured carefully. Gayer and Leider (1957) report a single solubility value of 3.0 x 10^{-6} mol U/1000 gm H_2O (0.7 ppm) for $U(OH)_4$ at 25°, and have shown that the solubility of $U(OH)_4$ increases only by about a factor of 2 in alkaline solutions at pH values up to 13. Miller (1958) found

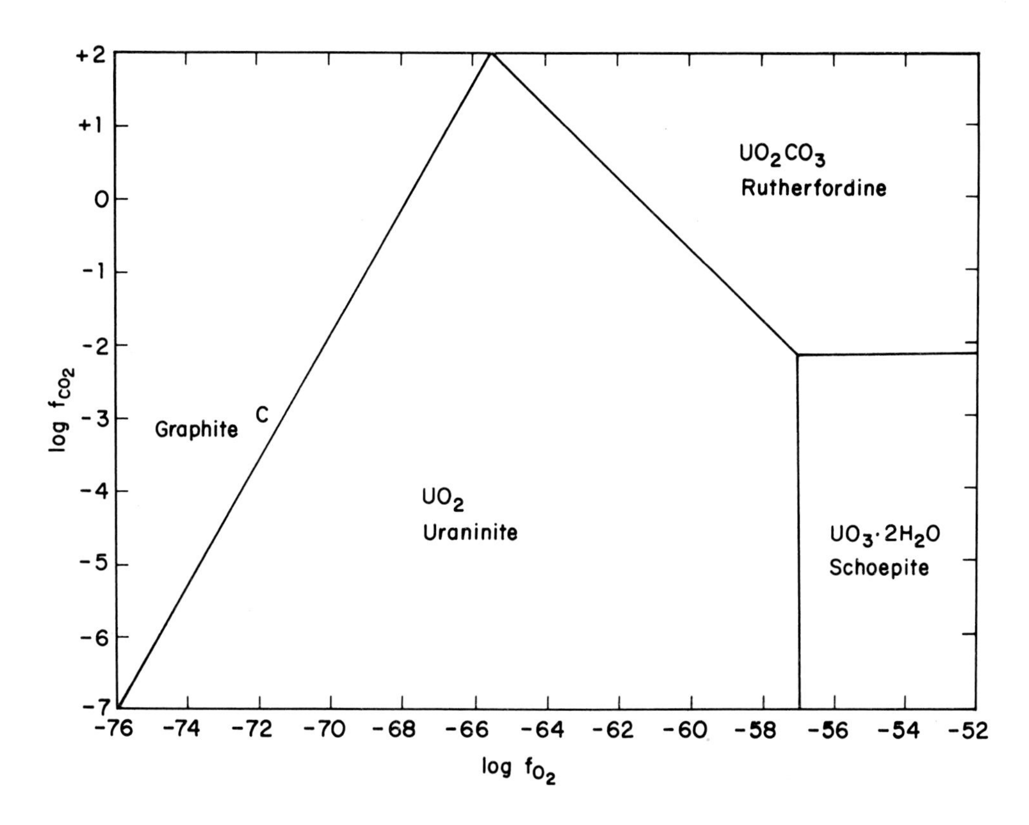

Fig. 4-2. Log f_{O_2} -log f_{CO_2} diagram for relevant solid phases in the system U-O-C-H at 25° C.

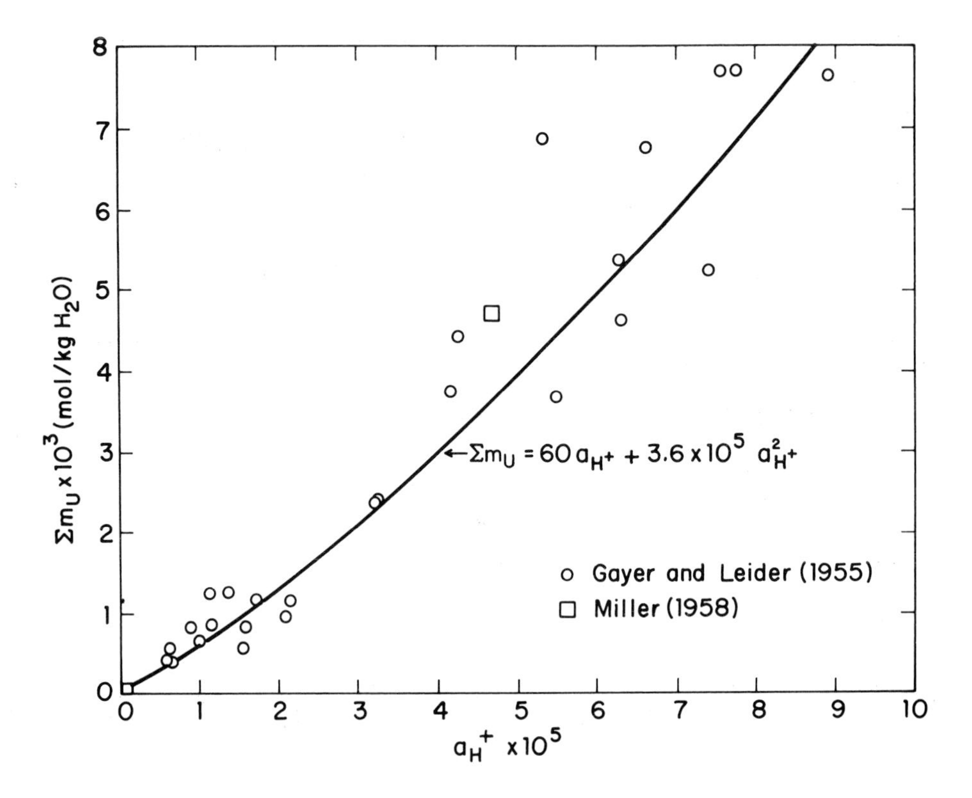

Fig. 4-3. Concentration of uranium in solutions saturated with respect to $UO_3 \cdot H_2O$ at 25° C (data from Gayer and Leider, 1955; Miller, 1958).

concentrations of less than 10^{-6} mol U/liter (<0.2 ppm) after reduction of $UO_2(CO_3)_3^{-4}$ solutions by H_2S at pH = 8, T = 25°C, and after reduction of $UO_2F_6^{-4}$ solutions by H_2S below 100°C.

In acid media the concentration of uranium in solution increases rapidly. Miller (1958) reports a single solubility measurement of $10^{-2.8}$ mol U/liter (ca. 700 ppm) for pitchblende in equilibrium with H_2S at 25°C and a pH of 2.8, but he states that this solubility may be too high due to the persistence of metastability at low temperatures. A rapid increase in $U(OH)_4$ solubility with decreasing pH was reported by Gayer and Leider (1957), but direct comparison of these data with those of Miller (1958) is impossible, because the final pH values in their experiments were not given.

The most extensive recent measurements of the solubility of uraninite are those of Lemoine (1975). Lemoine's solubility data for uraninite in water are shown in Figure 4-4. At 20°C the average of two rather uncertain measurements gives a concentration of 3-13 ppm uranium (1-5 x 10^{-5} mol U/1000gm solution). This value is almost certainly much larger than the solubility of $U(OH)_4$ at near-neutral pH values and low f_{O_2}; however, it agrees quite well with the solubility of $UO_3 \cdot H_2O$ and of $UO_3 \cdot 2\ H_2O$ described above, suggesting that Lemoine's runs were in equilibrium with uraninite and a hydrated form of UO_3. Support for this interpretation is found in the presence of small, pale yellow hexagonal crystals in the two run products which Lemoine examined optically. If a uranyl phase was present, then Lemoine's runs were not buffered by the intended magnetite-hematite or magnetite-hematite-pyrite buffer assemblages, but rather by the assemblage UO_2-uranyl hydrate. Two additional arguments speak against the efficacy of Lemoine's intended f_{O_2} buffers. Firstly, his reported uraninite solubilities were the same at a given temperature and pressure both for runs which included oxygen buffers and those that did not. Secondly, it is extremely unlikely that the magnetite-hematite and magnetite-hematite-pyrite assemblages can act as f_{O_2} buffers via hydrogen diffusion through a platinum wall at the low temperatures and relatively short run times of Lemoine's experiments. It is much more likely that UO_2 reacted with water to produce a small amount of UO_3 hydrates and hydrogen, so that the f_{O_2} values of Lemoine's runs were controlled by the presence of two uranium phases. If this is true, then Lemoine observed the solubility of uraninite at the upper f_{O_2} limit of its stability field, where its solubility is identical to that of UO_3 or the UO_3 hydrates.

The solubility data for UO_2 and $UO_3 \cdot 2H_2O$ near 25°C are summarized in preliminary fashion in Figure 4-5. At high values of f_{O_2} the uranium concentration in solution is probably controlled by the solubility of schoepite ($UO_3 \cdot 2H_2O$). The solubility of this phase increases rapidly in the direction of greater acidity, but is rather insensitive to pH on the alkaline side. At values of f_{O_2} below the uraninite-schoepite boundary the uranium concentrations in solutions saturated with respect to UO_2 are probably much lower than those in equilibrium with schoepite for a given pH. The estimated value of 10^{-6} (<1 ppm) for Σm_U at a pH of 3 and a temperature of 25°C (Hostetler and Garrels, 1962) is included in Figure 4-5 to illustrate this effect. At present the transition region is poorly defined, but it is known that the concentration of uranium in solution increases with increasing f_{O_2} in this region, because the concentration of U^{+6} complexes is increased by an increase of f_{O_2}. If uraninite at equilibrium with schoepite has the

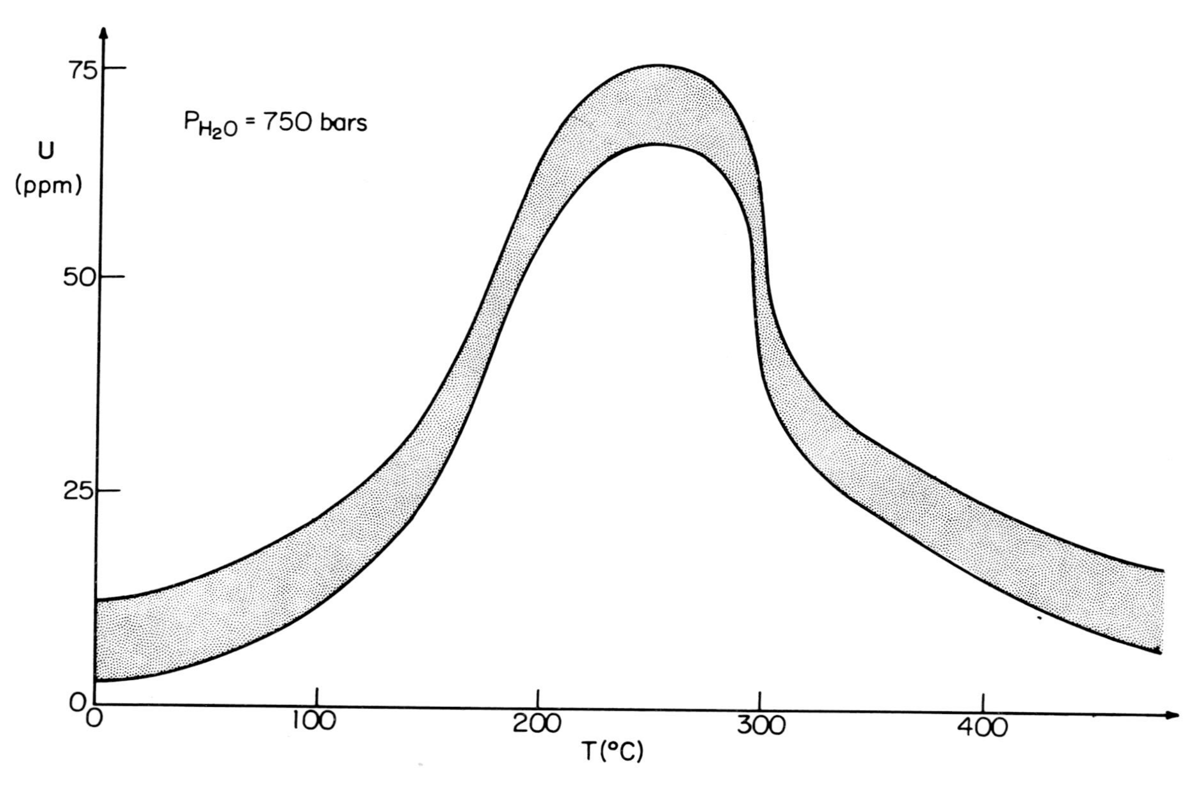

Fig. 4-4. Solubility of UO_2 in water as a function of temperature; f_{O_2} in the experimental runs was probably buffered by the equilibrium between UO_2 and hydrous or anhydrous UO_3 (redrawn from Lemoine, 1975).

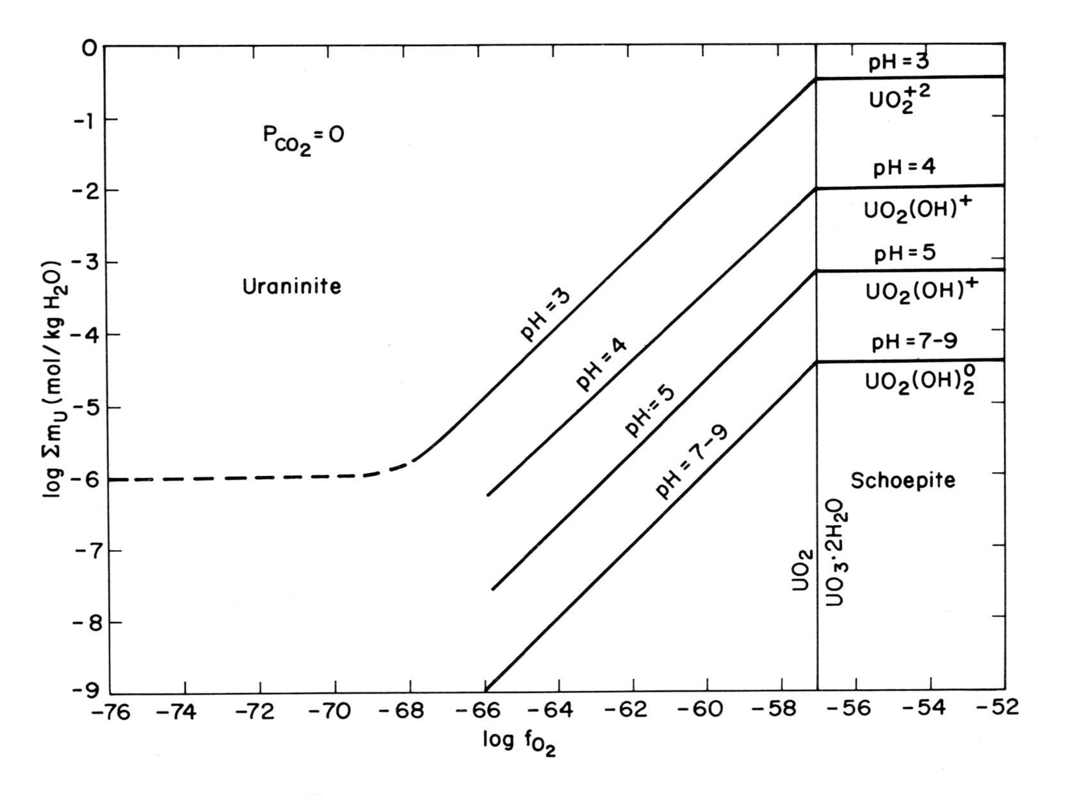

Fig. 4-5. Concentration of uranium in solutions in equilibrium with uraninite and schoepite at 25° C as a function of f_{O_2} and pH.

composition UO_2, then

$$UO_2 + ½O_2 + H^+ \rightleftharpoons UO_2(OH)^+ \quad (4\text{-}1)$$

and

$$K_{4-1} = \frac{a_{UO_2(OH)^+}}{f_{O_2}^{½} \cdot a_{H^+}}$$

The activity of $a_{UO_2(OH)^+}$ and of all similar U^{+6} complexes is therefore proportional to $f_{O_2}^{½}$ at constant pH. Figure 4-5 was constructed on this basis. If the composition of uraninite at 25°C is a strong function of f_{O_2} and approaches $UO_{2.6}$ at the uraninite-schoepite boundary, the uranium concentration is a somewhat different function of f_{O_2}. The data of Lemoine (1975) are currently the most complete for temperatures above 25°C. In pure water at near-neutral pH the concentration of uranium in solutions equilibrated with uraninite (and probably UO_3 and/or one of its hydrates) passes through a maximum near 260°C. The few measurements of Miller (1958) are in need of confirmation, but they do suggest that the increase in uranium concentration with decreasing pH becomes much less pronounced toward higher temperatures. For example, at a pH of 2 the solubility of uraninite (at a poorly defined f_{O_2} value) decreases rapidly between 100 and 200°C.

The data of Rafal'skiy ind Osipov (1967) for solutions containng sulfate and sulfides and saturated with respect to ZnS, (Zn,Fe)S, $CuFeS_2$, or PbS indicate that the uranium concentration in equilibrium with uraninite decreases with increasing temperature between 200 and 360°C. The uranium concentration in this temperature range lies between $10^{-4.7}$ and $10^{-6.7}$ mol/1 (approximately 5 to 0.05 ppm respectively) and decreases at a given temperature in the sequence low-iron sphalerite, marmatite, chalcopyrite, galena.

5. Uranium concentrations in aqueous solutions saturated with respect to minerals in the system U-O-H_2O-CO_2

Rutherfordine, UO_2CO_3, is probably the only uranyl carbonate mineral of importance. The CO_2 pressure in a vapor phase coexisting with rutherfordine and schoepite or its dehydration products has been measured by Sergeyeva et al. (1972), who also report the solubility of rutherfordine in equilibrium with approximately 1 atm CO_2 as a function of pH and temperature. Their data are plotted in Figure 4-6. There is clearly very little difference between the solubility of rutherfordine at 25 and 50°C.

In the low-pH region the solubility of UO_2CO_3 increases with decreasing pH. The slope of -1.5 that represents this effect is almost certainly due to the presence of both $UO_2(OH)^+$ and of UO_2^{+2} in solution. More detailed measurements would probably show a region with a slope of -1 merging as pH decreases with a region of slope -2.

Between a pH of 4.4 and 5.6 the solubility of UO_2CO_3 is independent of pH, reflecting the presence of a neutral complex, presumably $UO_2CO_3^\circ$ or one or more of its hydrates. The concentration of uranium in this pH region is $10^{-4.3}$ mol U/1000 gm H_2O

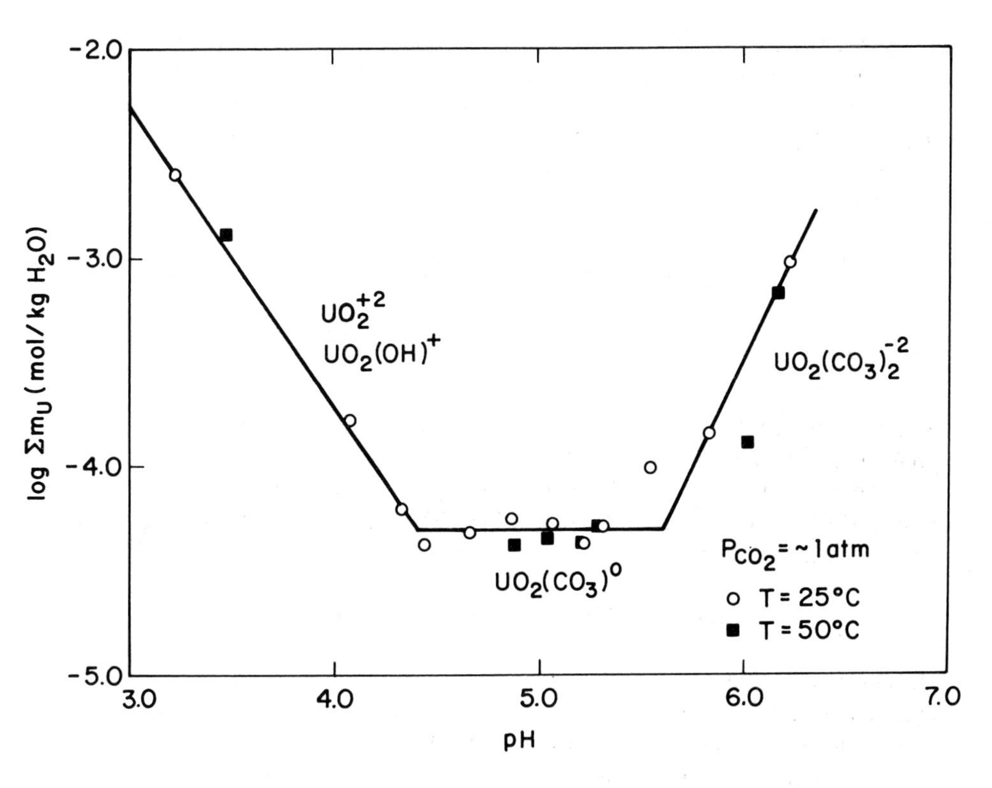

Fig. 4-6. Uranium concentration as a function of pH in solutions saturated with respect to rutherfordine ($UO_2 \cdot CO_3$) at 25 and 50°C and a CO_2 pressure of approximately 1 atmosphere (data from Sergeyeva et al., 1972).

(12 ppm). This rather high value is the minimum uranium content in solutions equilibrated with rutherfordine at 25°C and a CO_2 pressure of 1 atm.

At pH values in excess of 5.4 the uranium content rises with a slope of +2. Such a slope is readily explained by the presence of the uranyl dicarbonate complex (UDC), $UO_2(CO_3)_2^{-2}$. In equilibrium with rutherfordine the activity of this complex is controlled by the reaction

$$UO_2CO_3 + CO_2 + H_2O \leftrightharpoons UO_2(CO_3)_2^{-2} + 2H^+ \qquad (4\text{-}2)$$

$$K_{4\text{-}2} = \frac{a_{UO_2(CO_3)_2^{-2}} \cdot a_{H^+}^2}{P_{CO_2}} \cong 10^{-15.5}$$

Sergeyeva et al. (1972) have made a good case for accepting this value for K_{4-2} in preference to the value proposed by McClaine et al. (1958).

The dominance of the UDC complex in Na_2CO_3 solutions was suggested by the results of Blake et al. (1956) who found that UO_2CO_3 dissolves in a 1:1 ratio with Na_2CO_3 by the reaction

$$UO_2CO_3 + Na_2CO_3 \rightarrow UO_2(CO_3)_2^{-2} + 2Na^+ \qquad (4\text{-}3)$$

Dissolution of uranyl carbonate in sodium bicarbonate solutions resulted in immediate evolution of carbon dioxide and, eventually, in solutions with compositions similar to those in the system $UO_2(CO_3)$ — Na_2CO_3 — H_2O.

The uranyl tricarbonate complex (UTC) becomes an important species only at relatively high CO_3^{-2} activities. Blake et al. (1956) found that the equilibrium constant K'_{4-4} for the reaction

$$UO_2(CO_3)_3^{-4} \rightleftharpoons UO_2(CO_3)_2^{-2} + CO_3^{-2} \qquad (4\text{-}4)$$

$$K'_{4-4} = \frac{m_{UDC} \cdot m_{CO_3^{-2}}}{m_{UTC}} \cong 3 \times 10^{-4}$$

in solutions of ionic strength of about 2. This value of K'_{4-4} is in reasonable agreement with the value of the equilibrium constant K_{4-4} proposed by Garrels and Hostetler (1962):

$$K_{4-4} = \frac{a_{UDC} \cdot a_{CO_3^{-2}}}{a_{UTC}} \cong 10^{-3.8}$$

The experiments of Sergeyeva et al. (1972) indicate that the concentration of the two complexes should be equal at a pH of about 7.2. Because their measurements only extend up to a pH of 6.2, it is not surprising that their solubility data reflect only the effect of the UDC complex on the solubility of rutherfordine to the right of the solubility minimum.

Lemoine (1975) and Naumov (1961) have studied the concentration of uranium in $NaHCO_3$ solutions at temperatures in the hydrothermal range. Lemoine's starting material was UO_2; Naumov's was UO_3. Their results are roughly comparable. Both found uranium concentrations in excess of 10,000 ppm at low temperatures, and a rapid decrease in uranium concentration with increasing temperature. In the absence of positive identifications of run products and data regarding the composition of their final solutions, the results of these experiments are difficult to evaluate. It seems likely, however, that UO_2 in Lemoine's experiments reacted with H_2O to form H_2 and UO_3, which then reacted with HCO_3^- to form UDC by the reaction

$$UO_3 + 2HCO_3^- \rightarrow UO_2(CO_3)_2^{-2} + H_2O$$

The presence of yellow hexagonal crystals in his run products is in agreement with this interpretation. The observed increase in solubility with run time indicates that equilibrium was probably not reached in Lemoine's experiments. This could well have been due to the slowness of the UO_2-H_2O reaction, or possibly, at high temperatures, to the slow loss of H_2 through the capsule walls. Sergeyeva et al. (1972) suggest that the run products in the experiments of Naumov (1961) were probably sodium uranates. A great deal of work is still needed before it will be possible to define the solubility of the minerals in the system U-O-H_2O-CO_2 at temperatures above 50°C.

Figure 4-7 summarizes the solubility of UO_2 and UO_2CO_3 in solutions at 25°C at a CO_2 pressure of 1 atm. The data for the rutherfordine field are quite well defined. Within this field uranium concentrations pass through a minimum with increasing pH (see Figure 4-6). In the UO_2 stability field uranium concentrations in solution at a given pH decrease in proportion to $f_{O_2}^{1/2}$ provided the composition of uraninite is not markedly affected by f_{O_2}. Because U^{+4} apparently does not form carbonate complexes, the minimum uranium content in solutions at a given pH is the same as shown in Figure 4-5. However, before these minimum solubilities are reached, the solution becomes saturated with respect to graphite. The graphite boundary defines the lowest value of f_{O_2} that can be reached stably by these solutions at 25°C and in equilibrium with 1 atm CO_2.

6. The effect of other components on the concentration of uranium in aqueous solutions

In addition to the normal ionic strength effects, and those of the components discussed above, the concentration of uranium in aqueous solutions is influenced most by the presence of additional complexing agents. Among these the most important are probably sulfate (SO_4^{-2}) and fluoride (F^-).

Uranyl ions form the sulfate complexes $UO_2(SO_4)_2^{-2}$ and $UO_2(SO_4)_3^{-4}$; the uranous ion forms the complexes USO_4^{+2}, $U(SO_4)_3^{-2}$, and $U(SO_4)_4^{-4}$. Marshall (1955) determined the solubility of $UO_3 \cdot H_2O$ in sulfuric acid at temperatures between 150° and 290°C. UO_3/H_2SO_4 ratios were in excess of 0.6 even at the highest temperatures and in 1×10^{-3} m H_2SO_4 solutions. $UO_3 \cdot H_2O$ is therefore reasonably soluble in H_2SO_4

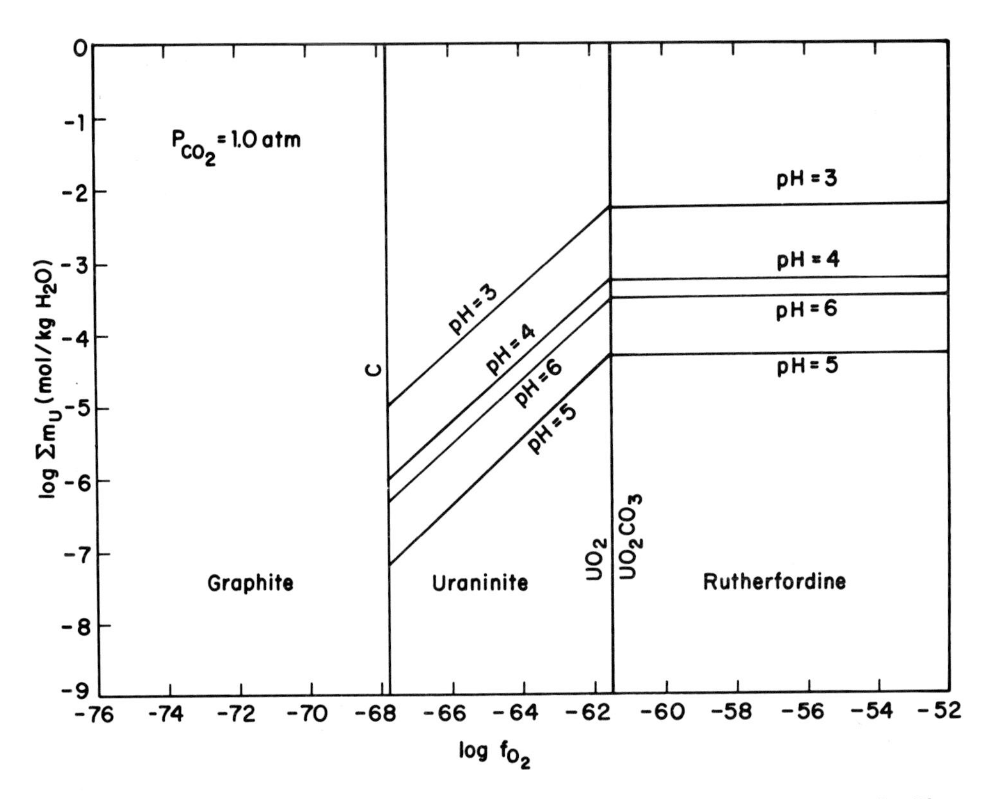

Fig. 4-7. Concentration of uranium as a function of f_{O_2} in solutions saturated with respect to uraninite and/or rutherfordine at 25° C, a CO_2 pressure of 1 atm, and an ionic strength of about 0.02.

throughout the range of hydrothermal ore deposit formation. However, the quench pH of Marshall's solutions at 25°C were nearly all below 4.0, and not enough is known about complexing in the range of his experiments to project his data into the higher, geologically more interesting near-neutral pH range. The uranyl ion forms several complexes with F^- containing up to six fluoride ions; $UO_2F_6^{-4}$ is stable up to a pH of 6.7 at 25°C. Uranous ions form a soluble fluoride complex, possibly UF_2^{+2}, which is stable up to a pH of 4. There is some indication that all of these complexes become less stable toward higher temperatures. Furthermore, their apparent instability in the pH range that is probably characteristic of most hydrothermal solutions suggests that the influence of fluoride, and probably sulfate, complexes on the transport of uranium is not of first order importance. Additional research, however, is needed to determine more accurately the effect of these complexes on uranium solubility.

7. Summary and conclusions

Despite the paucity of definitive data for the stability and solubility of the important uranium minerals, a few conclusions of considerable geologic importance can be drawn:

1. Natural uraninite spans nearly the entire compositional range between UO_2 and U_3O_8. Unfortunately the phase relations in the U-O system at temperatures below 400°C are not well known; this complicates the interpretation of solubility data.
2. At 25°C the stability field of uraninite is bordered on the high f_{O_2} side by the stability field of UO_3 or, in the presence of water, by the stability field of either $UO_3 \cdot 2H_2O$ (schoepite) or $UO_3 \cdot H_2O$.
3. At CO_2 pressures in excess of $10^{-2.2}$ atm rutherfordine ($UO_2 \cdot CO_3$) is stable at room temperature. Toward higher temperatures progressively larger CO_2 pressures are required to stabilize this mineral.
4. Solubility measurements of hydrated UO_3 at 25°C have shown that in acid solutions hexavalent uranium is present largely as $UO_2(OH)^+$ and UO_2^{+2}.
5. In the system U-O-H_2O-HCl the concentration of uranium in solution saturated with respect to $UO_3 \cdot H_2O$ or $UO_3 \cdot 2H_2O$ is about 10 ppm in neutral solutions, rises rapidly with decreasing pH, but increases only very slowly with rising pH.
6. The solubility of UO_2 at low values of f_{O_2}, where only tetravalent U ions are stable, is probably much less than 1×10^{-6} molar (0.2 ppm) at near-neutral values of pH. Within the stability field of uraninite the concentration of uranium therefore rises considerably toward higher values of f_{O_2} until the UO_2-$UO_3 \cdot 2H_2O$ boundary is reached.
7. In the presence of 1 atm CO_2 rutherfordine is stable at 50°C. The concentration of uranium in solutions saturated with respect to rutherfordine between 25 and 50°C passes through a minimum between pH 4.4 and 5.6. In solutions with $pH \leq 4.4$ U^{+6} is present largely as $UO_2(OH)^+$ and UO_2^{+2}, in solutions with pH between 5.6 and 7.2 largely as $UO_2(CO_3)_2^{-2}$ (UDC), and in solutions with pH values in excess of ca. 7.2 largely as $UO_2(CO_3)_3^{-4}$ (UTC). The importance of these carbonate complexes for the transport of uranium in oxidizing solutions is considerable at low temperatures and probably at high temperatures as well.
8. Sulfate and fluoride complexes of uranium may be of importance in uranium transport, but their effect is probably less pronounced than that of the O^{-2}, OH^-, and CO_3^{-2} complexes.

9. With increasing temperature the solubility of UO_2 in equilibrium with UO_3 probably passes through a maximum near 260°C. There is a rapid decrease in the solubility of UO_2 below 200°C and probably also above 300°C.

10. The presence of $NaHCO_3$ increases the concentration of uranium in solutions equilibrated with UO_2 and UO_3 throughout the hydrothermal range, although the uranium concentration in solutions of a given $NaHCO_3$ concentration seems to decrease rapidly with increasing temperature above 25°C.

References for Chapter 4

Ackermann, R.J. and Chang, A.T., 1973, Thermodynamic characterization of the U_3O_8 phase: Jour. Chem. Thermodynamics, *5*, 873-890

Affortit, C., 1969, Chaleur spécifique de UN, UC, et UO_2: High Temperature -High Pressure, *1*, 27-33.

Blake, C.A., Coleman, C.F., Brown, K.B., Hill, D.G., Lowrie, R.S. and Schmitt, J.M., 1956, Studies in the carbonate-uranium system: Jour. Amer. Chem. Soc., *78*, 5978-5983.

Fitzgibbon, G.C., Pavone, D. and Holley, C.E., Jr., 1967, Enthalpies of solution and formation of some uranium oxides: Jour. Chem. Eng. Data, *12*, 122-125.

Fredrickson, D.R. and Chasanov, M.G., 1970, Enthalpy of uranium dioxide and sapphire to 1500°K by drop calorimetry: Jour. Chem. Thermodynamics, *2*, 623-629.

Frondel, C., 1958, Systematic mineralogy of uranium and thorium: U.S. Geol. Survey Bull. 1064, 400 pp.

Garrels, R.M., 1955, Some thermodynamic relations among the uranium oxides and their relation to the oxidation states of the uranium ores of the Colorado Plateau: Am. Mineralogist, *40*, 1004-1021.

Garrels, R.M. and Christ, C.L., 1959, Behavior of uranium minerals during oxidation: U.S. Geol. Survey Prof. Paper 320, 81-89.

Garrels, R.M. and Larsen, E.S., 3rd, 1959, Geochemistry and mineralogy of the Colorado Plateau uranium ores: U.S. Geol. Survey Prof. Paper 320, 236 pp.

Garrels, R.M. and Pommer, A.M., 1959, Some quantitative aspects of the oxidation and reduction of the ores: U.S. Geol. Survey Prof. Paper 320, 157-164.

Garrels, R.M. and Richter, D.H., 1955, Is carbon dioxide an ore-forming fluid under shallow-earth conditions?: Econ. Geology, *50*, 447-458.

Gayer, K.H. and Leider, H., 1955, The solubility of uranium trioxide, $UO_3 \cdot H_2O$, in solutions of sodium hydroxide and perchloric acid at 25°C: Jour. Amer. Chem. Soc., *77*, 1448-1450.

Gayer, K.H. and Leider, H., 1957, The solubility of uranium (IV) hydroxide in solutions of sodium hydroxide and perchloric acid at 25°C: Canadian Jour. Chem., *35*, 5-7.

Gronvold, F., Kveseth, N.J., Sveen, A. and Tichy, J., 1970, Thermodynamics of the UO_{2+x} phase. I, Heat capacities of $UO_{2.017}$ and $UO_{2.254}$ from 300 to 1000°K and electronic contributions: Jour. Chem. Thermodynamics, *2*, 665-679.

Hostetler, P.B. and Garrels, R.M., 1962, Transportation and precipitation of uranium and vanadium at low temperatures, with special reference to sandstone-type uranium deposits: Econ. Geology, *57*, 137-167.

Leitnaker, J.M. and Godfrey, T.G., 1966, Thermodynamic properties of uranium carbides via the U-C-O system: Jour. Chem. Eng. Data, *11*, 392-394.

Lemoine, A., 1975, Contribution a l'étude du comportement de UO_2 en milieu aqueux a haute température et haute pression: unpub. doctoral thesis, Nancy, 111 pp.

Marshall, W.L., 1955, Simplified high temperature sampling and use of pH for solubility determinations, system uranium trioxide-sulfuric acid-water: Anal. Chem., *27*, 1923-1927

McClaine, L.A., Bullwinkel, E.P. and Huggins, J.C., 1956, Chemistry of the carbonate uranium compounds, theory and applications: Internat. Conf. Peaceful Uses of Atomic Energy, *8*, 29-42, United Nations.

McKelvey, V.E., Everhart, D.L. and Garrels, R.M., 1955, Origin of uranium deposits: Econ. Geology, 50th Anniversary Volume, 464-533.

Miller, L.J., 1958, The chemical environment of pitchblende: Econ. Geology, *53*, 521-545.

Naumov, G.B., 1961, Some physicochemical characteristics of the behavior of uranium in hydrothermal solutions: Geochemistry, No. 2, 127-147.

Pério, P., 1953, L'oxydation de UO_2 à basse température: Bull. Soc. Chim., 256-263.

Rafal'skiy, R.P. and Osipov, B.S., 1967, Equilibria in systems containing uranium and sulfides of heavy metals at 200-360°C: Geochem. International, *4*, 202-213.

Rand, M.H. and Kubaschewski, O., 1963, *Thermochemical Properties of Uranium Compounds:* Interscience, New York.

Ricci, J.E. and Loprest, F.J., 1955, Phase relations in the system sodium oxide-uranium trioxide-water at 50 and 75°C: Jour. Amer. Chem. Soc., *77*, 2119-2129.

Sergeyeva, E.I., Nikitin, A.A., Khodakovskiy, I.L. and Naumov, G.B., 1972, Experimental investigation of equilibria in the system UO_3-CO_2-H_2O in 25-200°C temperature interval, Geochem. International, *9*, 900-910.

Chapter 5
THE DEPOSITION OF PITCHBLENDE FROM HYDROTHERMAL FLUIDS

1. Introduction

The data in Chapter 4 are obviously incomplete, but they tend to confirm the conclusion that the amount of uranium that can be carried in oxidizing aqueous solutions is more than sufficient to account for the formation of hydrothermal uranium deposits. In acid solutions $UO_2(OH)^+$ and UO_2^{+2} are probably the major aqueous species of uranium. $UO_2(OH)_2^\circ$ and $UO_2CO_3^\circ$ are important in near-neutral solutions, and in the alkaline range $UO_2(CO_3)_2^{-2}$ and $UO_2(CO_3)_3^{-4}$ predominate. On the other hand, the solubility of UO_2 is very low in solutions which are sufficiently reducing so that uranyl complexes are essentially absent. The reduction of U^{+6} to U^{+4} is therefore a likely mechanism for the deposition of pitchblende from hydrothermal solutions. This mechanism has been used successfully to explain the primary mineralogy of the uranium deposits of the Colorado Plateau (e.g. Garrels and Larsen, 1959). Other mechanisms for precipitating pitchblende have also been proposed. G.B. Naumov and B.P. Poty (see for instance Naumov and Mironova, 1969a, 1969b; Tugarinov and Naumov, 1974; Poty et al., 1974) have proposed that CO_2 loss from ore solutions during boiling has played a major role in the precipitation of pitchblende in hydrothermal uranium deposits. In addition, it is possible that mechanisms which have been routinely invoked to explain the deposition of ore minerals in base and precious metal deposits have played a role in the formation of uranium deposits. These include changes in the temperature and total pressure, and changes in the pH and cation ratios due to fluid-wall rock reactions.

2. Reduction as a mechanism for pitchblende deposition

Historically, the great attraction of the reduction mechanism for explaining the origin of sandstone-type uranium and uranium-vanadium deposits of the Colorado Plateau has rested on two readily observed facts: (1) the highly oxidized state of the rocks through which the ore-forming fluids must have passed, and (2) the presence of carbon trash and plant remains which served as the requisite reducing agents for U^{+6}. Redox reactions are also of obvious importance in the formation of sedimentary roll-type deposits such as those of the Shirley Basin in Wyoming, although the reducing agents responsible for precipitating pitchblende in this type of deposit are less easily identified.

In many hydrothermal uranium deposits hematite is commonly associated in time and space with pitchblende. Three-quarters of the 41 hydrothermal uranium deposits discussed in Chapter 2 contain hematite either as a vein or a wall rock alteration mineral (see Tables 2-2 and 2-3). In most of these deposits pitchblende is reported to have been deposited with or after hematite. The common occurrence of hydrothermal hematite is distinctly unusual. Most hydrothermal base and precious metal deposits contain very little if any hematite. Furthermore, the base metal stages of hydrothermal uranium deposits are typically free of hematite.

The uraninite-schoepite boundary at 25°C lies well within the stability field of hematite; even at elevated temperatures, where schoepite is replaced by $UO_3 \cdot H_2O$ and finally by UO_3 as the stable U^{+6} oxide in the presence of water, this is almost certainly true. Uranium in hydrothermal solutions which are in equilibrium with hematite is therefore largely present as a constituent of uranyl complexes. The concentration of these complexes in aqueous solutions can certainly be large enough to account for the formation of known hydrothermal uranium deposits. The concentration of base metals in hydrothermal ore-forming fluids is probably on the order of 10-5000 ppm (Czamanske et al., 1963; Roberts, 1975). The largest base metal deposits contain some 10 to 100 times the tonnage of metal in the largest hydrothermal uranium deposits. If the total volumes of water involved in the production of these ore deposits are comparable, then uranium concentrations between 0.1 and 500 ppm are likely in the solutions which give rise to major hydrothermal uranium deposits. These concentrations are 10^3 to 10^5 times greater than those in average surface and ground waters, but they are by no means impossible to attain either in mildly acid or in mildly alkaline oxidizing solutions (see Chapter 4).

Likewise, several lines of field evidence suggest the operation of a reduction mechanism for pitchblende deposition in hydrothermal uranium deposits. For example, in a number of districts throughout the world the distribution of hydrothermal uranium mineralization shows a clear preference for the more reducing rock units of the region. One of the better examples of this relation are the uranium deposits of Golden Gate Canyon in the Colorado Front Range. In their description of these deposits Adams and Stugard (1956) point out that virtually all of the pitchblende-bearing veins of the area occur where faults cut two narrow bands of hornblende gneiss that are found in a dominantly biotite gneiss terrane. What is significant about this example is that the hornblende gneiss unit contains almost four times as much reduced iron (Fe^{2+}) as the surrounding biotite gneiss. Furthermore, a comparison of fresh and hydrothermally altered hornblende gneiss samples shows that, although total iron remains essentially constant in both rock types, most of the Fe^{2+} has been oxidized to Fe^{3+} (occurring as hematite) in the altered samples (Adams and Stugard, 1956, Figure 48). Inasmuch as the hematitic alteration is known to be part of the uranium mineralization event for the Golden Gate Canyon deposits, the evidence strongly suggests that the precipitation of pitchblende results from the reduction of U^{+6} to U^{+4} by the oxidation of the reduced iron in the wall rock adjacent to ore fluid channelways. It should be emphasized at this point that in other hydrothermal uranium deposits, reduced S (sulfide) or C (graphite, organic matter, etc.) may also effectively serve as the reducing agent causing pitchblende deposition; ferrous iron is by no means the only important reductant controlling hydrothermal uranium mineralization.

Pitchblende deposition frequently occurs with or before the deposition of pyrite or marcasite in hydrothermal uranium deposits. In the group of 41 deposits whose paragenetic sequences were summarized in Table 2-3, pitchblende was deposited before or with FeS_2 in 28 (68%). Furthermore, base metal sulfide stages followed the pitchblende stage in 80% of the 41 deposits. Typically then, more reducing conditions follow the highly oxidizing conditions which characterize the early hematite stage of hydrothermal uranium deposits. Pitchblende is usually deposited between the hematite and sulfide stages, presumably in response to the reduction of U^{+6} to U^{+4} in the ore fluid. In the hydrothermal deposits at Echo Bay, N.W.T., Canada, a quartz-hematite stage was followed by a pitchblende-hematite stage, then an arsenide stage, and finally a sulfide stage (Figure 5-1). Robinson and Ohmoto (1973) suggest that the hydrothermal solutions of this deposit evolved along a path from A to G in the log f_{O_2} - log f_{S_2} diagram shown in Figure 5-2. Along such a path sulfate is gradually reduced to sulfide (see for instance Raymahashay and Holland, 1969). Robinson and Ohmoto (1973) have demonstrated a dramatic progressive change with time in the isotopic composition of sulfur in sulfide minerals at Echo Bay (Table 5-l). This change of δS^{34}, from ca. -22 ‰ (CDT) in the earliest sulfide to ca. $+27$ ‰ (CDT) in the last sulfide, is certainly consistent with, and probably best explained by, a gradual reduction of sulfate to sulfide in solutions with an approximately constant initial δS^{34} value of $+25 \pm 3$ ‰ (CDT). Robinson and Ohmoto (1973) tentatively attribute the cause of the progressive reduction to reaction of the hydrothermal solutions with ferromagnesian minerals; however, reduction through mixing with reducing gases is not ruled out. Although the data of Robinson and Ohmoto (1973) indicate only small changes in the isotopic composition of carbon in the ore forming fluids at Echo Bay, Sassano et al. (1972) found a range of 0 to -17 ‰ (PDB) for δC^{13} in carbonates from the Beaverlodge district, Saskatchewan (Table 5-2). Such a large variation in δC^{13} implies either a major change in the source of carbon during the evolution of the hydrothermal system or the operation of redox reactions involving graphite and/or hydrocarbons. Reduction of sulfate to sulfide and U^{+6} to U^{+4} could therefore have involved the oxidation of graphite or carbon compounds in addition to, or instead of, Fe^{+2}.

Extremely large variations in the isotopic composition of sulfur in sulfides from hydrothermal uranium deposits are not restricted to Echo Bay. Heyse (1971) found a range of δS^{34} from -37 ‰ to $+4$ ‰ (CDT) for the sulfides of the Schwartzwalder mine, Colorado (Table 5-1). The δS^{34} values of several other deposits listed in Table 5-1 also show variations which are large, though not quite as spectacular.

Additional studies, such as that of Robinson and Ohmoto (1973), which carefully relate isotopic data to the paragenesis of hydrothermal uranium deposits are needed in order to determine whether the interpretation developed for the Echo Bay data is generally applicable. If so, the accepted chemical model for the origin of sandstone-type uranium deposits is applicable to hydrothermal uranium deposits without major changes.

3. CO_2 loss as a mechanism of pitchblende precipitation

Poty et al. (1974) and Naumov and Mironova (1969a, 1969b) have shown that fluid inclusions in hydrothermal uranium deposits frequently contain more than 1 mole per cent CO_2. At Limousin (France) there exists a clear relation between uranium ore grade and the CO_2 concentration in associated inclusion fluids; in addition, the

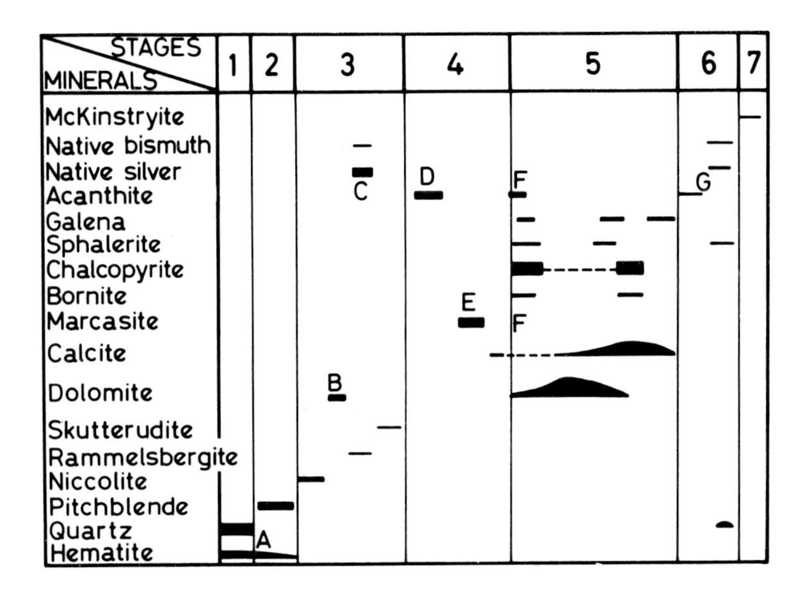

Fig. 5-1. Paragenesis of vein minerals, Echo Bay mine, Great Bear Lake district, Canada. Points A to G represent characteristic mineral phase(s) considered in the geochemical model shown in Fig. 5-2: A = hematite; B = first appearance of dolomite; C = first appearance of native silver; D = first appearance of acanthite; E = appearance of marcasite; F = first appearance of the mineral assemblage, acanthite + chalcopyrite + carbonates; G = change from acanthite to native silver (from Robinson and Ohmoto, 1973).

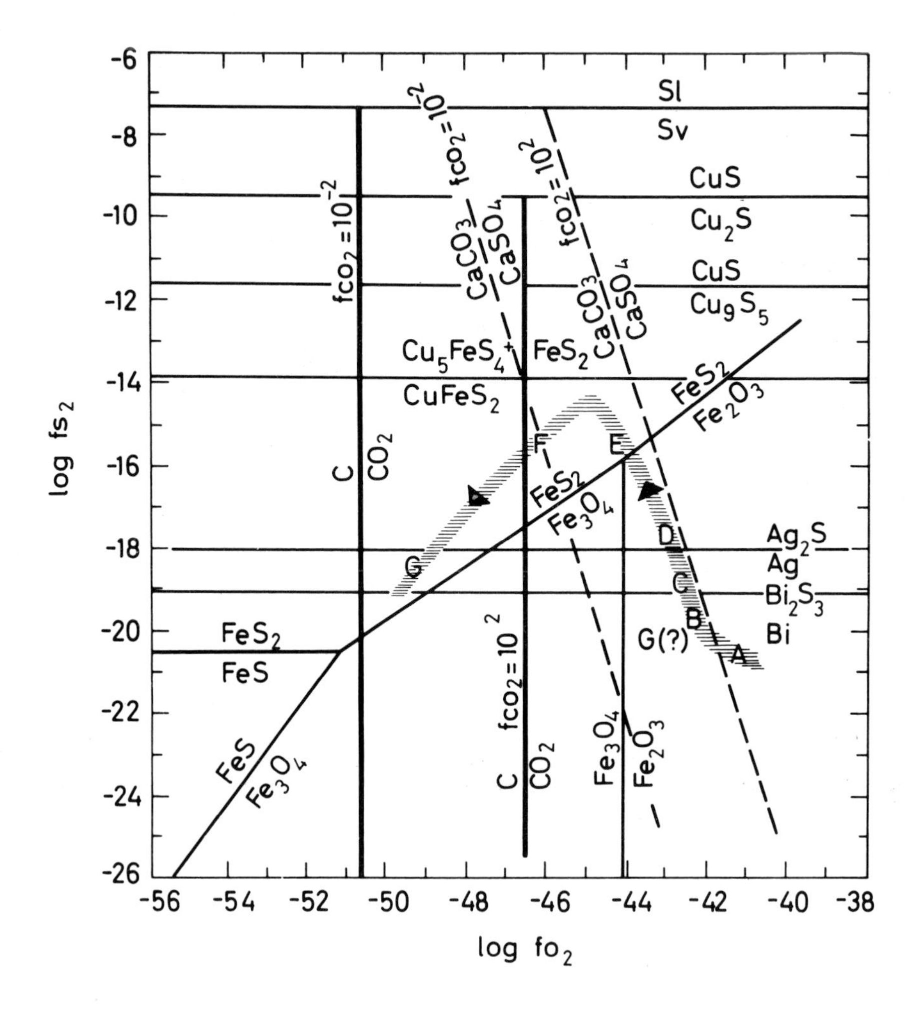

Fig. 5-2. Log f_{S_2}—log f_{O_2} mineral stability relationships at 150°C. Points A to G correspond to those in Fig. 5-1; path A to G_{II} is the probable path for the evolution of ore fluids at the Echo Bay mine, N.W.T., Canada (from Robinson and Ohmoto, 1973).

Table 5-1. Sulfur Isotope Data for Hydrothermal Uranium Deposits

Locality	Mineral(s)	Paragenetic Stage*	δS^{34}(‰CDT)	Geothermometry	Reference
Great Bear Lake district, Canada:					
Echo Bay mine:	Host rock pyrite	Pre-vein	+2.4 to +5.1	200± 30°C (stage 5)	Robinson & Ohmoto (1973)
	Pyrite of amphibole-magnetite veins	Pre-vein	+2.1 to +2.2		
	Acanthite	4	−21.5 to −19.7		
	Marcasite	4	−2.0 to +0.4		
	Acanthite	5	−4.1 to −0.5		
	Galena	5	+2.4 to +19.9		
	Sphalerite	5	+4.1 to +12.6		
	Chalcopyrite	5	+3.7 to +13.5		
	Bornite	5	+5.1 to +14.6		
	Acanthite	6	+21.3		
	Mckinstryite	7	+23.0 to +27.1		
Terra mine and associated deposits:	Chalcopyrite	Pre-vein	−10.2 to +4.0	~200°C	Robinson & Badham (1974)
	Pyrite	Pre-vein	−9.6 to +5.1		
	Chalcopyrite	1**	+1.0 to +3.9		
	Marcasite	1**	+0.9 to +4.0		
	Chalcopyrite	3a	+1.4 to +2.1		
	Galena	3a	−2.0 to +4.2		
	Chalcopyrite	3b	−4.2 to +4.0		
	Galena	3b	−8.0 to +1.5		
	Pyrite	3b	−1.2		
	Bismuthinite	3b	+2.7		
	Matildite	3b	+1.4 to +1.9		
	Chalcopyrite	4	+4.0 to +4.3		
	Pyrite	4	−26.0***		
United States:					
Orphan mine, Arizona:	Pyrite	-	−27.2 to −2.8		Gornitz & Kerr (1970)
	Pyrite + chalcopyrite	-	−19.0 to −11.7		
	Chalcopyrite	-	−2.8		
	Chalcocite	-	−18.6		
	Sphalerite	-	−19.0		
	Sphalerite + galena	-	−22.6		
	Galena	-	−21.2		
Front Range, Colorado:					
Schwartzwalder mine:	Pyrite	-	−36.98 to +0.80	-	Heyse (1971)
	Pyrite + chalcopyrite	-	−33.82		
	Pyrite + chalcocite	-	+1.33		
	Pyrrhotite	-	−2.41 to −1.30		
	Chalcopyrite	-	+1.39		
	Galena	-	−5.33		
	Chalcocite + bornite	-	+3.96		
Central City district:	-	-	−2 to +4	-	Jensen (1967)
Marysvale district, Utah:					
Prospector mine:	-	-	−3 to +5	-	Jensen (1967)
Freedom mine:	-	-	−4 to −3	-	
Farmer John mine:	-	-	−3 to −2	-	
Deer Trail mine:	-	-	0 to +3	-	
South Alligator River district, Australia:	Pyrite	-	−5.6 to +12.3	-	Ayres & Eadington (1975)
	Galena	-	−3.4 to +10.4		
Shinkolobwe deposit, Zaire:	Pyrite + siegenite	-	−1.3 to +6.9	-	Dechow & Jensen (1965)

* For paragenetic diagrams of individual deposits see Part II of this book.
** Pitchblende stage.
*** Average of several analyses.

Table 5-2. Carbon Isotope Data for Hydrothermal Uranium Deposits

Locality	Mineral	Paragenetic Stage*	δC^{13} (‰ PDB)	Reference
Beaverlodge District:				
Fay mine/Bolger pit	Calcite	a_0	−6.73 to −2.45	Sassano et al. (1972)
	Calcite	A**	−8.67 to −1.60	
	Dolomite	B**	−5.24 to +0.55	
	Calcite	C	−3.77 to −1.62	
	Calcite	D**	−17.04 to −4.28	
	Calcite	E	−16.82 to −15.88	
Great Bear Lake District:				
Echo Bay mine	Dolomite	3	−4.1 to −2.8	Robinson and Ohmoto (1973)
	Calcite	4	−3.9 to −2.3	
	Dolomite	5	−3.0 to −1.4	
Terra mine and associated deposits	Dolomite	Pre-vein	−6.1	Robinson and Badham (1974)
	Dolomite	1**	−4.9 to −4.2	
	Dolomite	2a	−4.5 to −2.8	
	Calcite	2a	−9.2 to −3.8	
	Dolomite	2b	−4.2 to −2.6	
	Dolomite	3a	−4.6 to −2.5	
	Calcite	3a	−3.7	
	Calcite	3b	−4.0	
	Dolomite	3b	−4.5 to −3.1	
	Calcite	4	−7.3 to −4.7	

* For paragenetic diagrams of individual deposits, see deposit descriptions given in Part II of this book.
** Pitchblende-depositing stage.

hydrothermal fluid was boiling at the time of trapping of the Limousin fluid inclusions (Poty et al., 1974). In the Erzgebirge, fluid inclusions in minerals which were deposited just before pitchblende contain especially high concentrations of CO_2. On the other hand, minerals deposited immediately after pitchblende contain considerably lower CO_2 contents (Naumov, 1973; Tugarinov and Naumov, 1974). Such observations led to the conclusion that pitchblende precipitated from highly carbonated solutions due to a decrease in their CO_2 content. However, as the following discussion shows, the loss of CO_2 from hydrothermal solutions is somewhat unlikely to result in the precipitation of pitchblende.

If, as seems likely, uranium is transported largely as the uranyl dicarbonate (UDC) complex in CO_2 -rich hydrothermal fluids, then the reaction expressing equilibrium between the fluid and pitchblende of composition $UO_{2.0}$ is

$$UO_2 + \tfrac{1}{2}O_2 + 2CO_2 + H_2O \rightleftharpoons UO_2(CO_3)_2^{-2} + 2H^+ \qquad (5\text{-}1)$$

$$K_{5-1} = \frac{a_{UO_2(CO_3)_2^{-2}} \cdot a^2_{H^+}}{f_{H_2O} \cdot a^2_{CO_2} \cdot f_O^{1/2}}$$

It is evident that a decrease of the CO_2 activity in a solution saturated with respect to UO_2 at constant pH and f_{O_2} will reduce the activity of $UO_2(CO_3)_2^{-2}$, and that pitchblende will tend to precipitate. However, fairly special conditions have to prevail in order to keep the pH and f_{O_2} constant during CO_2 loss from solution. A loss of CO_2 implies a loss of acid. The solution will therefore tend to become more alkaline during CO_2 loss. This, in turn, will decrease, or even prevent the further precipitation of UO_2. The nature of this effect is perhaps best illustrated by the reaction

$$UO_2 + \tfrac{1}{2}O_2 + 2\,HCO_3^- \rightleftharpoons UO_2(CO_3)_2^{-2} + H_2O \qquad (5\text{-}2)$$

for which

$$K_{5-2} = \frac{a_{UO_2(CO_3)_2^{-2}} \cdot f_{H_2O}}{f_{O_2}^{1/2} \cdot a^2_{HCO_3^-}}$$

This demonstrates that UDC is really a bicarbonate complex, and that the activity of $UO_2(CO_3)_2^{-2}$ in solutions saturated with pitchblende is proportional to $f_{O_2}^{1/2}$ and $a^2_{HCO_3^-}$. The behavior of $a_{HCO_3^-}$ during CO_2 loss due to the boiling of hydrothermal solutions depends on the compositions of the solution. In $NaHCO_3$-$KHCO_3$-CO_2-H_2O solutions, the concentration of HCO_3^- is essentially equal to the sum of the Na^+ and K^+ concentrations, and the addition or release of CO_2 has very little effect on $m_{HCO_3^-}$. On the other hand, in $CaCO_3$ - CO_2 - H_2O solutions CO_2 loss can result in the precipitation of $CaCO_3$ and a decrease in the HCO_3^- concentration via the reaction

$$Ca^{+2} + 2\,HCO_3^- \rightarrow CaCO_3 + CO_2 + H_2O \qquad (5\text{-}3)$$

However, pitchblende is not usually deposited simultaneously with calcite or other carbonates in hydrothermal uranium deposits; it is therefore unlikely that the simple release of CO_2 from hydrothermal solutions results in pitchblende deposition. Reaction of hydrothermal fluids with wall rocks will tend to buffer the solution pH during CO_2 release. If buffering is very effective, then CO_2 loss could well result in UO_2 precipitation; however, in many hydrothermal uranium deposits hydrogen ion metasomatism is weak and of very minor extent. It seems doubtful, therefore, that reaction with wall rocks has buffered pH during CO_2 loss.

Loss of CO_2 is apt to affect the oxidation state of the residual solutions. Although detailed calculations will have to be made to obtain precise measures of the various effects, it is likely that the f_{O_2} of the solutions will increase during boiling. The escape of CH_4, H_2 and H_2S will increase the oxidation state of the residual solution, but the loss of CO_2 will have the opposite effect. Since CH_4 is less soluble in aqueous solutions than CO_2, the loss of CH_4 to the vapor phase will tend to be proportionately greater than that of CO_2, and the net effect will probably be an increase in the oxidation state of the residual solutions.

It is therefore not obvious that CO_2 evolution moves hydrothermal solutions strongly in the direction of pitchblende precipitation. If this inference is borne out by a more thorough analysis, the main effect of the high CO_2 pressures observed in some uraniferous hydrothermal solutions is probably the stabilization of uranyl complexes via a proportionately higher bicarbonate concentration.

4. The effects of other parameters

The data of Lemoine (1975) show that in the system UO_2-UO_3-H_2O the concentration of uranium increases with temperature to ca. 260°C and then decreases rapidly. On the other hand the concentration of uranium in $NaHCO_3$ solutions appears to decrease monotonously with increasing temperature. The effect of temperature on the solubility of UO_2 in equilibrium with UO_3 and UO_3 hydrates is therefore a function of solution composition. Most hydrothermal uranium deposits have been formed below 250°C (see Chapter 3). A decrease in temperature can therefore either increase or decrease the solubility of UO_2, depending on the HCO_3^- concentration of the solution. Comparable data are not available for UO_2 solubility in acid solutions. The effect of other dissolved salts such as NaCl is probably small in the pH range where neutral complexes are dominant; but NaCl would increase the solubility of UO_2 and UO_3 hydrates both in the alkaline pH range, due to the effect of increasing ionic strength on the activity coefficients of the carbonate complexes, and in acid solutions, where $UO_2(OH)^+$ and UO_2^{+2} are dominant species.

Variations of total pressure have a rather modest effect on the solubility of most minerals in hydrothermal solutions. An increase of 1 kb in total pressure frequently produces an increase of about a factor of 2 in the solubility of slightly soluble compounds in hydrothermal solutions. The data of Lemoine (1975) suggest that the solubility of pitchblende is no exception to this rule. Total pressure effects are apt to be of importance only in P-T regions where the density of the solutions is a strong function of pressure.

pH has a profound effect on the solubility of pitchblende. At 25 and 50°C, a well defined pitchblende solubility minimum between pH 4.4 and 5.6 was documented by the experiments of Sergeyeva et al. (1972); this solubility minimum probably persists to higher temperatures (see Figure 4-6). The solubility behavior of pitchblende in water and in $NaHCO_3$ solutions suggests that the solubility minimum widens with increasing temperature, at least on the alkaline side.

5. Summary and conclusions

The classic explanation for the formation of sandstone-type uranium deposits also seems the best explanation for the formation of many hydrothermal uranium deposits. The presence of hematite as a vein and wall rock mineral in most hydrothermal uranium deposits is consistent with uranium transport in the +6 valence state. The deposition of sulfides and arsenides with or shortly after pitchblende is consistent with the reduction of U^{+6} to U^{+4} and the deposition of the very sparingly soluble UO_2 at low f_{O_2}. The isotopic composition of sulfur in sulfides and of carbon in carbonates observed in some hydrothermal uranium deposits, as well as the frequent localization of ore in the vicinity of reducing wall rocks, is consistent with, and probably best explained by, the reduction-depositon model.

The presence of rather large CO_2 concentrations in fluid inclusions from some hydrothermal uranium deposits suggests that CO_2 may be important for uranium transport, but the effect of CO_2 loss on UO_2 deposition is unclear. There are too few systematic data to define the effect of pH on the solubility of pitchblende at elevated temperatures and to evaluate the effect of pH changes on the development of hydrothermal uranium deposits, but at present it seems likely that the effects of changes in solution pH are not as important as those of redox reactions.

References for Chapter 5

Adams, J.W. and Stugard, F., Jr., 1956, Wall-rock control of certain pitchblende deposits in Golden Gate Canyon, Jefferson County, Colorado: U.S. Geol. Survey Bull. 1030-G, 187-209.

Ayres, D.E. and Eadington, P.J., 1975, Uranium mineralization in the South Alligator River Valley: Mineralium Deposita, *10*, 27-41

Czamanske, G.K., Roedder, E. and Burns, F.C., 1963, Neutron activation analysis of fluid inclusions for copper, manganese, and zinc: Science, *140*, 401-403.

Dechow, R. and Jensen, M.L., 1965, Sulfur isotopes of some Central African sulfide deposits: Econ. Geology, *60*, 894-941.

Garrels, R.M. and Larsen, E.S., 3rd, 1959, Geochemistry and mineralogy of the Colorado Plateau uranium ores: U.S. Geol. Survey Prof. Paper 320, 236 pp.

Gornitz, V. and Kerr, P.F., 1970, Uranium mineralization and alteration, Orphan mine, Grand Canyon, Arizona: Econ. Geology, *65*, 751-768.

Heyse, J.V., 1971, Mineralogy and paragenesis of the Schwartzwalder mine uranium ore, Jefferson County, Colorado: U.S. Atomic Energy Comm. Rept. GJO-912-1, 91 pp.

Jensen, M.L., 1967, Sulfur isotopes and mineral genesis: in Barnes, H.L., ed., *Geochemistry of Hydrothermal Ore Deposits*, 143-165, Holt, Rinehart and Winston, Inc., New York.

Lemoine, A., 1975, Contribution a l'étude du comportement de UO_2 en milieu aqueux a haute température et haute pression: unpub. doctoral thesis, Nancy, 111 pp.

Naumov, G.B., 1973, The behavior of the radioactive components in hydrothermal processes: in *Radioactive Elements in Rocks*, Izd-vo Nauka SO AN SSSR, Novosibirsk.

Naumov, G.B. and Mironova, O.F., 1969a, Migration of uranium in hydrothermal carbonate solutions (according to physicochemical data): in Khitarov, N.I., ed., *Problems of Geochemistry*, 166-175.

Naumov, G.B. and Mironova, O.F., 1969b, Das Verhalten der Kohlensäure in hydrothermalen Lösungen bei der Bildung der Quarz-Nasturan-Kalzit-Gänge des Erzgebirges: Zeitschr. Angew. Geologie, *15*, 240-241.

Poty, B.P., Leroy, J. and Cuney, M., 1974, Les inclusions fluides dans les minerais des gisements d'uranium intragranitiques du Limousin et du Forez (Massif Central, France): in *Formation of Uranium Deposits*, 569-582, Internat. Atomic Energy Agency, Vienna.

Raymahashay, B.C. and Holland, H.D., 1969, Redox reactions accompanying hydrothermal wall rock alteration: Econ. Geology, *64*, 291-305.

Roberts, S., 1975, Early hydrothermal alteration and mineralization in the Butte district, Montana: unpub. doctoral thesis, Harvard, 173 pp.

Robinson, B.W. and Badham, J.P.N., 1974, Stable isotope geochemistry and the origin of the Great Bear Lake silver deposits, N.W.T., Canada: Canadian Jour. Earth Sci., *11*, 698-711.

Robinson, B.W. and Ohmoto, H., 1973, Mineralogy, fluid inclusions, and stable isotopes of the Echo Bay U-Ni-Ag-Cu deposits, Northwest Territories, Canada: Econ. Geology, *68*, 635-656.

Sassano, G.P., Fritz, P. and Morton, R.D., 1972, Paragenesis and isotopic composition of some gangue minerals from the uranium deposits of Eldorado, Saskatchewan: Canadian Jour. Earth Sci., *9*, 141-157.

Sergeyeva, E.I., Nikitin, A.A., Khodakovskiy, I.L. and Naumov, G.B., 1972, Experimental investigation of equilibria in the system UO_3-CO_2-H_2O in 25-200°C temperature interval: Geochem. International, *9*, 900-910.

Tugarinov, A.I. and Naumov, G.B., 1974, Die Migrations- und Absatzverhältnisse des Urans bei der endogenen Erzbildung: Zeitschr. Angew. Geologie, *20*, 410-413.

Chapter 6
ORIGIN OF HYDROTHERMAL URANIUM DEPOSITS

1. Introduction

The previous two chapters have made a strong case for the hypothesis that uranium is transported predominantly as U^{+6} complexes in hydrothermal solutions and precipitated as pitchblende in hydrothermal deposits following the reduction of uranium to the tetravalent state. This chapter examines the geologic settings where uranium leaching, transport and deposition are most likely to occur, compares these settings with the observed environments of hydrothermal uranium deposits, and offers some general guidelines for finding hydrothermal uranium deposits.

Four ingredients seem to be required for the formation of hydrothermal uranium deposits:

(1) Highly oxidized solutions.
(2) Source(s) of leachable uranium.
(3) Reducing agent(s) at the site of pitchblende deposition.
(4) A suitable hydrologic setting in which the above ingredients can operate. In this respect, to form an ore deposit the system must allow a sufficiently large volume of uraniferous solution to be channeled through a sufficiently small volume of rock favorable to pitchblende deposition. The four ingredients are discussed in order below.

2. Sources of highly oxidized solutions in the subsurface

The solubility data in Chapter 4 are incomplete, but they do suggest that solutions whose oxidation states are well within the stability field of hematite are required for the efficient transport of uranium. There appear to be two classes of such solutions which may have participated in the formation of hydrothermal uranium deposits:

(1) Surface waters which have equilibrated with atmospheric oxygen and have maintained a high f_{O_2} in the subsurface, and
(2) Originally reducing subsurface solutions which have encountered and equilibrated with highly oxidized rocks.

Present-day surface waters which have equilibrated with atmospheric oxygen contain approximately 6 ml O_2 S.T.P./liter, i.e. a dissolved O_2 concentration of ca. 3 x 10^{-4} mol/liter. In such oxygenated solutions sulfur is present entirely as sulfate and sulfate complexes, nitrogen as N_2, nitrate and nitrate complexes, and carbon as dissolved CO_2, HCO_3^-, CO_3^{-2} and their complexes. The salinity of these solutions covers a broad range from rain water with only a few ppm dissolved salts to brines in evaporitic settings with salinities of several hundred thousand ppm.

The oxidation state of aqueous solutions in equilibrium with the atmosphere has always been determined by the oxygen content of the contemporaneous atmosphere. The history of atmospheric P_{O_2} is not well known, but various lines of evidence suggest that P_{O_2} was considerably lower prior to 2.0 b.y. ago (see for instance Holland, 1962, 1973), and that it has fluctuated about its present value during the last few hundred million years. Thus, rainwater, river water and seawater are probably more oxidizing now than they were in Archean time, and it has been suggested in an earlier chapter that this difference accounts for the absence of Archean hydrothermal uranium deposits. Caution, however, should be used in the application of this notion in prospecting. If, for example, the oxygen content of the atmosphere were reduced to 1% of its present level, the only major change in the chemistry of waters equilibrated with the atmosphere would be a reduction of the dissolved O_2 concentration to 1% of its present value; the oxidation state of sulfur, nitrogen, and carbon would remain virtually unchanged. However, such a decrease of the O_2 concentration would certainly reduce the ability of surface solutions to oxidize subsurface rocks; this in turn would surely reduce the likelihood, but might not eliminate the possibility, of generating hydrothermal uranium deposits by means of surface-derived solutions.

Although there is some uncertainty concerning the effect of f_{O_2} on the concentration of uranium in solutions saturated with respect to pitchblende at temperatures above 100°C, it is likely that the oxidation of originally reducing sub-surface waters can produce potential uranium ore solutions. Typical hydrothermal solutions contain enough reduced sulfur, largely as H_2S and HS^-, to place them well within the stability field of pyrite at f_{O_2} values below those of the magnetite-hematite boundary. Rock units which are sufficiently oxidized to raise the f_{O_2} values of normal hydrothermal solutions into the U^{+6} transporting range are not common, and red beds are the most likely candidates. Hematite-rich aquifers can increase the f_{O_2} of hydrothermal solutions until equilibrium with hematite has been reached, but the final value of f_{O_2} in these solutions depends on their composition prior to reaction with hematite (Raymahashay and Holland, 1969). For example, solutions containing large quantities of dissolved sulfate will tend to reach equilibrium with hematite at a higher f_{O_2} than solutions containing little or no dissolved sulfate. Most red beds have at least some constituents which can raise the f_{O_2} of contained fluids well above the f_{O_2} value at the magnetite-hematite boundary. Calcite together with gypsum or anhydrite in red beds is not uncommon. It has been shown (see for instance Holland, 1965) that at geologically reasonable CO_2 pressures solutions saturated with respect to these minerals can have f_{O_2} values well within the hematite stability field (Figure 5-1). Pyrolusite, MnO_2, is also found in red beds. The Mn_3O_4-MnO_2 boundary lies at such a high f_{O_2} value that MnO_2 is a very attractive oxidant for initially reducing hydrothermal solutions; in the presence of sufficient quantities of MnO_2 such solutions should become highly efficient carriers of U^{+6}.

The isotopic composition of hydrogen and oxygen are the best indicators of the past history of hydrothermal solutions. Unfortunately there are at present no data for the isotopic composition of hydrogen in waters from which pitchblende was precipitated in hydrothermal uranium deposits; the available δO^{18} values for hydrothermal uranium deposits are listed in Table 6-1. With the exception of the data of Sassano et al. (1972) for the Fay mine and the Bolger Pit in the Beaverlodge district, Saskatchewan, all of the calculated δO^{18} values for the ore fluids of hydrothermal uranium deposits are from the

Great Bear Lake district, and these approximate the value of present-day seawater. Robinson and Ohmoto (1973) suggest on the basis of sulfur and oxygen isotope data that circulating seawater was involved in the deposition of the ores at Echo Bay; if this is correct, then the very high salinity of some of the fluid inclusions in this deposit (see Table 3-2) indicates that during ore deposition a process occurring either at the surface or in the subsurface strongly modified the salinity of this seawater. The calculated δO^{18} values do not exclude the possibility that the Great Bear Lake ore fluids were of meteoric origin; however, the data do seem to rule out the equilibration of either meteoric water or seawater with silicate rocks at magmatic temperatures prior to ore depositon. On the other hand, the data of Sassano et al. (1972) for the Beaverlodge district suggest that such an equilibration may well have taken place. The fluid inclusion data of Pagel (1976) and Pagel and Jaffrezic (1976, written communication) for the Rabbit Lake deposit in Saskatchewan suggest that at least some of the ore fluids were probably waters resulting from the diagenesis of the nearby (and originally overlying) Athabasca sandstone. Unfortunately, present fluid inclusion and isotopic data are too incomplete to give more than vague hints regarding the history of uranium ore-forming fluids. Detailed studies of hydrothermal uranium deposits are clearly needed; particularly fruitful results are apt to be obtained from studies of geologically young deposits where convincing data are available for the isotopic composition of contemporary seawater and local rainwater.

3. Sources of leachable uranium

In order to evaluate the favorability of a potential uranium source rock, it is important to consider at least two factors:

1. In what form is the uranium contained in the rock? Is it contained in a mineral from which it is relatively easily released (e.g. uraninite, biotite and hornblende), is it in a refractory mineral like zircon or apatite, or is the uranium loosely held interstitially?
2. Is the rock composition such that a hydrothermal fluid in equilibrium with it would be likely to leach and transport uranium?

Granitic rocks appear to be a particularly favorable source of leachable uranium, because they are often enriched in uranium (>5 ppm), and because the quantity of reducing minerals is usually small. Barbier et al. (1967) report the presence of accessory uraninite in the two-mica granites of the hydrothermal uranium districts of the Limousin and Vendée regions of France. Furthermore, the experimental studies of Szalay and Samsoni (1969) and Larsen et al. (1956) have demonstrated that hexavalent uranium is readily leached from granitic and rhyolitic rocks by dilute acid solutions. At least half of the uranium content of granitic rocks is usually leachable even during surficial weathering (Barbier, 1974; Barbier and Ranchin, 1969). Granitic rocks are closely associated with most hydrothermal uranium deposits (see Table 2-1), and uranium-rich granitic rocks are present in many hydrothermal uranium districts (e.g. Beaverlodge, the Midnite mine, the Massif Central, and the Colorado Front Range). The general favorability of granite as a source rock does not, of course, imply that the scavenging of uranium from other uranium-rich rock types is necessarily unimportant (see Chapter 1).

Table 6-1. Oxygen Isotope Data for Hydrothermal Uranium Deposits

Locality	Mineral	Paragenetic Stage*	δO^{18} (⁰/00 SMOW)	Geothermometry	Calculated δO^{18} of ore fluid (⁰/00 SMOW)	Reference
Beaverlodge District: Fay mine/Bolger pit	Vein carbonates	a_o	+5 to +7	120-190°C (stage?)	+7 ± 1 (all stages)	Sassano et al. (1972); Robinson (1955)
	" "	A**	+8 to +12			
	" "	B**	+10 to +14			
	" "	C	+8 to +11			
	" "	D**	+10 to +18			
	" "	E	+13 to +15			
Great Bear Lake District: Echo Bay mine	Dolomite	3	+21.9 to +22.9		+1 ± 2.5 (all stages)	Robinson & Ohmoto (1973)
	Calcite	4	+16.1 to +17.4			
	Dolomite	5	+11.8 to +15.1			
Eldorado mine	Quartz	1**	+18.0	135 ± 15°C (stage 1)	+0.3 (stage 1)	Clayton & Epstein (1958); Taylor (1967); Robinson and Ohmoto (1973)
	Hematite	1**	−2.8			
	Calcite	5	+12.8			
Terra mine and associated deposits	Dolomite	1**	+13.9 to +14.1	˜200°C (all stages)	+2 (all stages)	Robinson and Badham (1974)
	Dolomite	2a	+15.3 to +17.2			
	Calcite	2a	+7.0 to +9.7			
	Dolomite	2b	+14.1 to +16.9			
	Dolomite	3a	+12.6 to +22.0			
	Calcite	3a	+18.2			
	Calcite	3b	+14.2			
	Dolomite	3b	+13.1 to +15.5			
	Calcite	4	+7.8 to +10.8			

* For paragenetic diagrams of individual deposits, see deposit descriptions given in Part II of this book.
** Pitchblende-depositing stage.

If a solution which has equilibrated with the present-day atmosphere is allowed to react with UO_2 in the subsurface so that all of the initial dissolved O_2 is used to convert UO_2 to UO_3, and if all the UO_3 produced by this reaction enters the solution, then the total uranium addition to the solution will be ca. 6 x 10^{-4} mol/1 (140 ppm). This value is, of course, an extreme upper limit, because most of the dissolved oxygen will almost certainly be removed from the solution during its passage through the subsurface by the oxidation of Fe^{+2}, S^{-2}, and C^{o}. However, ground waters can also scavenge uranium which has previously been oxidized to U^{+6}, for example in a soil zone. The ease with which hexavalent uranium is leached from granitic rocks during normal surficial weathering processes suggests that the uranium content of potential ore solutions is not necessarily limited by the capacity of subsurface fluids to oxidize and dissolve U^{+4}. Nevertheless, the formation of large hydrothermal uranium deposits probably requires something above and beyond such leaching, because surface and shallow ground waters rarely contain more than a few ppb uranium, and such low concentrations would require the passage of improbably large volumes of waters through the ore system.

4. Reducing agents

Ferrous iron (Fe^{+2}), sulfide (S^{-2}) and reduced carbon are quantitatively the most important reducing agents in the upper crust. Biotite, hornblende, pyrite, graphite, and hydrocarbons are capable of removing molecular oxygen efficiently from surface and subsurface solutions and reducing U^{+6} to U^{+4}. That this process can lead to the precipitation of pitchblende is suggested by the many areas in which uranium mineralization is spatially related to rocks containing large amounts of these reductants. For example, in several hydrothermal uranium deposits of the Massif Central, France, the richest uranium ore is localized at the intersection of vein structures and minette (potassium feldspar-biotite lamprophyre) dikes (Sarcia et al., 1958). In both the Erzgebirge region (Ruzicka, 1971) and the Great Bear Lake district (Robinson and Badham, 1974) pitchblende mineralization is concentrated in wall rocks containing abundant pre-vein sulfides. This relation is particularly well exhibited in the Johanngeorgenstadt district (Viebig, 1905). Dodson et al. (1974) have demonstrated a clear lithologic control for the hydrothermal uranium deposits of the Darwin region, Northern Territory, Australia. There, all of the important uranium deposits in three widely separated districts are found in the same carbonaceous and chloritic pelitic rock unit. At the Midnite mine, Washington, uranium ore occurs in sulfidic and locally carbonaceous metasediments adjacent to a granitic intrusive; the intrusive is essentially devoid of hydrothermal pitchblende mineralization (Nash and Lehrman, 1975). Similarly, the occurrence of uranium mineralization in some skarn zones (e.g. at Mary Kathleen, Australia, and at Traversella, Italy) may well be due to the relatively reducing nature of skarn minerals such as hedenbergite (Burt, 1972).

In some hydrothermal uranium deposits, however, no obvious reducing agents can be identified. In the Schwartzwalder mine, for example, some pitchblende ore is found in apparently clean quartzites as well as in rocks containing potential reductants for U^{+6}. In several of the French deposits no redox abnormalities have been found in the two-mica granites which host the uranium mineralization. At Marysvale, Utah, no obvious reducing agents for U^{+6} have been observed. It is possible that in some of these districts a limited amount of reductants in the country rocks enabled oxidizing, uraniferous waters to travel considerable distances before being reduced sufficiently to precipitate pitchblende.

It is also possible that in hydrothermal uranium deposits such as those mentioned in the preceding paragraph, U^{+6} has been reduced by H_2S, H_2 or hydrocarbons in solutions with which the more oxidizing, uranium-rich waters mixed in the sub-surface. Gaseous and liquid hydrocarbons have been observed in fluid inclusions from Mississippi Valley Pb-Zn-fluorite deposits (see for example Hall and Friedman, 1963), and fluid mixing has been proposed for the origin of these deposits (Anderson, 1975; Beales, 1975). In this regard it is interesting to note that Cuney (1974) and Poty et al. (1974) report the occurrence of hydrocarbon-bearing fluids during the deposition of certain vein stages at the Bois-Noirs (Limouzat) uranium deposit in France. In addition, Kranz (1968) reports that there are hydrocarbons in fluid inclusions in uraniferous fluorite from the Oberpfalz district, Germany; these hydrocarbons may have been responsible for the reduction of all or part of the uranium in this area.

5. Hydrologic settings

Hydrothermal solutions have traditionally been considered to have originated at depth. This explanation was acceptable as long as the classic magmatic-hydrothermal hypothesis offered the only reasonable explanation for the formation of hydrothermal ore deposits. The strong isotopic evidence which has accumulated during the past decade in favor of the involvement of surface waters in the development of hydrothermal ore deposits implies that hydrothermal solutions can circulate in convective cells, and that ore deposition is not necessarily confined to the ascending limb of such cells. Large convection cells may be several kilometers in diameter.

The formation of hydrothermal uranium deposits "per descensum" has in recent years been strongly advocated in France (see for instance Moreau et al., 1966; Barbier, 1974), where many uranium veins of shallow vertical extent seem to be related to the Hercynian erosion surface. Langford (1974), Knipping (1974) and Derry (1973), among others, have applied meteoric water models to the genesis of hydrothermal uranium deposits in other parts of the world. Smith (1974) has recently summarized the evidence for and against a supergene origin for hydrothermal uranium deposits.

Published fluid inclusion evidence (see Chapter 3) for several hydrothermal uranium deposits precludes a very shallow origin involving rain water or dilute ground waters (see for instance Poty et al., 1974; Kotov et al., 1968, 1970; Tugarinov and Naumov, 1969). Additional fluid inclusion data are needed to prove or disprove a shallow supergene origin for the many other hydrothermal uranium deposits to which this model has been applied; however, the absence of data confirming a supergene origin does not rule out the possibility that pitchblende deposition does take place from descending fluids of different origins at considerable depths. It is important here to distinguish ore formation by shallow supergene processes from the more general concept of ore deposition by descending fluids at any depth.

There are several criteria for distinguishing ore deposition during fluid descent from deposition during fluid ascent, but these may be difficult to apply in practice:

(1) If the flow of hydrothermal fluids is too rapid for fluid-wall rock thermal equilibration, the temperature of ascending solutions should be higher than that of the adjacent wall rocks; therefore alteration sequences should normally develop in a negative temperature gradient outward from veins. The reverse would be true for descending solutions. Unfortunately, wall rocks surrounding hydrothermal uranium veins are usually only weakly altered, and where alteration is strong the means for determining the direction of temperature gradients are not yet available.

(2) Quartz is normally deposited when hydrothermal fluid flow takes place in the direction of decreasing quartz solubility. Because the solubility of quartz decreases with falling temperature and pressure, quartz should be deposited in veins where solution flow was upward, and not in veins where the solutions flowed downward. Unfortunately, this simple criterion may be invalid where silica is released during wall rock alteration, because quartz may then appear as a gangue mineral even if flow is in the direction of higher quartz solubility.

(3) The geometry of vein systems might be of help in deciding between ore deposition by descending and ascending solutions, but as far as we know this criterion has not yet been applied to hydrothermal uranium deposits. What is clear, however, is that at least some hydrothermal uranium deposits, such as the Fay-Verna mine in the Beaverlodge district, were demonstrably formed over a depth range of more than 1000 m, and that others were formed at temperatures which are difficult to reconcile with a very shallow origin except in regions of exceptionally strong geothermal gradients.

(4) The sense of asymmetric crystal growth and crustification in pitchblende veins could, in principle, be used to determine the direction of fluid flow.

(5) The shape of observed zoning contours based on uranium grade, element ratios, alteration types, etc. could in some cases indicate the direction of fluid flow.

The following features should characterize hydrothermal uranium deposits formed by shallow supergene processes and should be useful for distinguishing such deposits from those formed by other processes:

1) Relatively limited vertical extent.
2) Close spatial association with a major erosional surface or unconformity.
3) Mineralogic, fluid inclusion, and isotopic data should indicate a low temperature and pressure of formation.
4) Fluid inclusion data should indicate that the ore-forming fluids were dilute aqueous solutions with negligible CO_2 contents.
5) Vein mineralogy in general should be simple, but this depends to a great extent on the composition of the source rocks being weathered and leached.
6) The age of the uranium ore deposit must not be greater than the age of the genetically related erosional surface or unconformity.
7) δD and δO^{18} data for inclusion fluids should indicate the involvement of meteoric water.
8) A local source for the vein constituents must be geochemically and/or

isotopically demonstrable. Of greatest importance, of course, is the presence of a weathered uranium-rich rock. δC^{13} and δS^{34} data, however, should indicate a local source for the carbon and sulfur of vein carbonates and sulfides.

9) Ore body morphology should reflect ore formation by a descending ore fluid. In this regard, it is interesting to note that many French uranium veins pinch out at a shallow depth. At the Midnite mine, Washington, ore body morphology also suggests a supergene origin; there, tongues of ore cross-cut metasediments without apparent regard for lithology as they extend downward and outward from a topographic high of uranium-rich intrusive granitic rocks (Nash and Lehrman, 1975).

Fluid flow in potential ore-forming environments is presumably driven either by mechanical forces, such as a difference of hydraulic head between the intake and discharge areas, or by the effect of thermal gradients on solution densities. Sufficient thermal gradients can certainly be produced by the presence of magmas or of recently emplaced intrusive or extrusive rocks. It would be interesting to determine whether thermal gradients resulting from a heterogeneous distribution of radioelements or from tectonic processes, such as large scale vertical displacements, could account for the necessary fluid flow. If a heterogeneous uranium distribution can by itself produce significant fluid circulation, then uranium deposit host rocks such as the uranium-rich granites in the French Massif Central could have served the double function of supplying uranium to the solutions and assuring its transport. If vertical displacements can generate sufficiently extensive circulation of subsurface fluids, then the location of hydrothermal uranium deposits such as the Schwartzwalder mine on the edge of the Colorado Front Range could be explained. The hydrology of such settings should be modeled carefully; until this is done the causes of fluid flow leading to the formation of hydrothermal uranium deposits will remain subject to speculation. Once they have been modeled, useful hydrologic constraints can be added to existing geochemical limits on genetic models.

6. Discussion and conclusions

The potential geologic settings of hydrothermal uranium deposits are limited by the chemical constraints which uranium oxidation, transport, and deposition impose on the origin and migration of the ore-forming fluids. The genetic options permitted within the bounds of the chemical constraints are summarized in Table 6-2. Three different generalized models for the formation of hydrothermal uranium deposits are implicit in Table 6-2; in all cases, pitchblende is precipitated through reduction of the f_{O_2} of the ore fluid:

1) **Supergene model:** very shallow uranium leaching by oxidizing surface waters, followed by pitchblende precipitation at depth.
2) **Deep meteoric water model:** uranium leaching by deeply circulating oxidizing waters of surface origin, followed by pitchblende deposition.
3) **Non-meteoric water model:** uranium leaching by deeply circulating oxidiz-

ing waters which are not of direct surface origin (e.g. magmatic, metamorphic and sedimentary formation waters), followed by pitchblende deposition.

It should be emphasized that the presence of red beds in the right hydrologic position is apparently required for the third model and would seem to be extremely useful in the second. In this regard, it is interesting that many important hydrothermal uranium deposits occur in crystalline basement rocks associated with, or overlain by, either red bed or clean sandstone units (e.g. the Beaverlodge district, Schwartzwalder mine, Rabbit Lake mine, Great Bear Lake district, and Cluff Lake deposit). Paleogeographic reconstructions suggest that other important hydrothermal uranium districts, such as the Erzgebirge, Príbram and the Massif Central may also have been overlain by continental red beds at the time of uranium mineralization.

Table 6-2. Permissible Options Leading to the Formation of Hydrothermal Uranium Deposits

I. Source of Oxidizing Solutions:

1. Surface waters (rain, rivers, oceans, evaporite brines).
2. Deep waters (magmatic, metamorphic, etc.) which have equilibrated with hematitic rocks.

II. Access of Oxidizing Solutions to Uranium Source Rocks (fluid flow may be upward, downward, or lateral):

1. Direct (only applicable to surface waters).
2. Via oxidizing or inert conduits (e.g. red beds or clean sandstones, or other non-reducing aquifers).

III. Uranium Transport from Source to Deposition Site (fluid flow may be upward, downward, or lateral):

1. Deposition directly adjacent to, or within, source rocks.
2. Deposition at a distance from source—transport via oxidizing or inert conduits (e.g. red beds or clean sandstones, or other non-reducing rocks).

IV. Reducing Agents for Uranium Deposition:

1. Reduction by solids containing Fe^{+2}, S^{-2}, or C^o.
2. Reduction by solutes and/or gases (Fe^{+2}, S^{-2}, H_2, hydrocarbons, etc.) following fluid mixing.

In summary, pitchblende is probably deposited in hydrothermal uranium deposits from oxidizing solutions in response to the reduction of U^{+6} to U^{+4}. The ore solutions may either be descending or deeply circulating surface waters which have equilibrated with atmospheric oxygen, or they may be deep waters of non-surface origin which have equilibrated with a hematitic aquifer. The uranium deposited in veins etc. was probably scavenged from uranium-rich rocks by the same oxidizing fluids. This model suggests that the present or past regional association of red beds with uranium-rich rocks (especially granites) may be a useful guide in the search for additional hydrothermal uranium deposits.

References for Chapter 6

Adams, J.W. and Stugard, F., Jr., 1956, Wall-rock control of certain pitchblende deposits in Golden Gate Canyon, Jefferson County, Colorado: U.S. Geol. Survey Bull. 1030-G, 187-209.

Anderson, G.M., 1975, Precipitation of Mississippi Valley-type ores: Econ. Geology, *70*, 937-942.

Barbier, J., 1974, Continental weathering as a possible origin of vein-type uranium deposits: Mineralium Deposita, *9*, 271-288.

Barbier, J. and Ranchin, G., 1969, Géochimie de l'uranium dans le Massif de Saint-Sylvestre (Limousin-Massif Central Français). Occurrences de l'uranium géochimique primaire et processus de remaniements: Sci. Terre Mém. 15, 115-157.

Barbier, J., Carrat, H.G. and Ranchin, G., 1967, Présence d'uraninite en tant que minéral accessoire usuel dans les granites à deux micas uranifères du Limousin et de la Vendée: Acad. Sci. Comptes Rendus, Ser. D, *264*, 2436-2439.

Beales, F.W., 1975, Precipitation mechanisms for Mississippi Valley-type ore deposits: Econ. Geology, *70*, 943-949.

Burt, D.M., 1972, Mineralogy and geochemistry of Ca-Fe-Si skarn deposits: unpub. doctoral thesis, Harvard, 256 pp.

Clayton, R.N. and Epstein, S., 1958, The relationship between O^{18}/O^{16} ratios in coexisting quartz, carbonate, and iron oxides from various geological deposits: Jour. Geology, *66*, 352-373.

Cuney, M., 1974, Le gisement uranifère des Bois-Noirs-Limouzat (Massif Central-France) — Relations entre minéraux et fluides: unpub. doctoral thesis, Nancy, 174 pp.

Derry, D.R., 1973, Ore deposition and contemporaneous surfaces: Econ. Geology, *68*, 1374-1380.

Dodson, R.G., Needham, R.S., Wilkes, P.G., Page, R.W., Smart, P.G. and Watchman, A.L., 1974, Uranium mineralization in the Rum Jungle — Alligator Rivers province, N.T., Australia: in *Formation of Uranium Ore Deposits*, 551-568, Internat. Atomic Energy Agency, Vienna.

Hall, W.E. and Friedman, I., 1963, Composition of fluid inclusions, Cave-in-Rock fluorite district, Illinois, and Upper Mississippi Valley Zn-Pb district: Econ. Geology, *58*, 886-911.

Holland, H.D., 1962, Model for the evolution of the earth's atmosphere: in Engel, A.E.J., James, H.L. and Leonard, B.F., eds., *Petrologic Studies* — A volume in honor of A.F. Buddington, 447-477, Geol. Soc. America.

Holland, H.D., 1965, Some applications of thermochemical data to problems of ore deposits II. Mineral assemblages and the composition of ore-forming fluids: Econ. Geology, *60*, 1101-1166.

Holland, H.D., 1973, Ocean water, nutrients and atmospheric oxygen: in Ingerson, E., ed., *Proceedings of Symposium of Hydrogeochemistry and Biogeochemistry, 1*, 68-81, The Clarke Co., Washington, D.C.

Knipping, H.D., 1974, The concepts of supergene versus hypogene emplacement of uranium at Rabbit Lake, Saskatchewan, Canada: in *Formation of Uranium Ore Deposits*, 531-549, Internat. Atomic Energy Agency, Vienna.

Kotov, Ye. I., Timofeev, A.V. and Khoteev, A.D., 1968, Formation temperatures of minerals of uranium hydrothermal deposits (abst.): Fluid Inclusion Research — Proceedings of C.O.F.F.I., *1*, 44.

Kotov, Ye. I. et al., 1970, Formation temperatures of some hydrothermal uranium deposits (abst.): Fluid Inclusion Research — Proceedings of C.O.F.F.I., *3*, 38.

Kranz, R.L., 1968, Participation of organic compounds in the transportation of ore metals in hydrothermal solutions: Inst. Mining and Metallurgy Trans., Section B, *77*, B26-B36.

Langford, F.F., 1974, A supergene origin for vein-type uranium ores in the light of the Western Australian calcrete-carnotite deposits: Econ. Geology, *69*, 516-526.

Larsen, E.S., Jr., Phair, G., Gottfried, D. and Smith, W.S., 1956, Uranium in magmatic differentiation: Internat. Conf. Peaceful Uses of Atomic Energy, *6*, 240-247, United Nations.

Moreau, M., Poughon, A., Puibaraud, . and Sanselme, H., 1966, L'uranium et les granites: Chronique Mines et Recherche Minière no. 350, 47-51.

Nash, J.T. and Lehrman, N., 1975, Geology of the Midnite uranium mine, Stevens County, Washington — A preliminary report (abst.): Geol. Soc. America Abs. with Programs, *7*, 634-635 (full text given in U.S. Geol. Survey Open-File Report 75-402, 36 pp.).

Pagel, M., 1976, Conditions de dépôt des quartz et dolomites automorphes du gisement uranifère de Rabbit Lake (Canada): 4e Réunion Annuelle des Sciences de la Terre, Paris (in press).

Pagel, M. and Jaffrezic, H., 1976, Analyses chimiques des saumures des inclusions du quartz et de la dolomite du gisement d'uranium de Rabbit Lake [Canada], unpublished manuscript.

Poty, B., Leroy, J. and Cuney, M., 1974, Les inclusions fluides dans les minerais des gisements d'uranium intragranitiques du Limousin et du Forez (Massif Central, France): in *Formation of Uranium Deposits*, 569-582, Internat. Atomic Energy Agency, Vienna.

Raymahashay, B.C. and Holland, H.D., 1969, Redox reactions accompanying wall rock alteration: Econ. Geology, *64*, 291-305.

Robinson, B.W. and Badham, J.P.N., 1974, Stable isotope geochemistry and the origin of the Great Bear Lake silver deposits, N.W.T., Canada: Canadian Jour. Earth Sci., *11*, 698-711.

Robinson, B.W. and Ohmoto, H., 1973, Mineralogy, fluid inclusions, and stable isotopes of the Echo Bay U-Ni-Ag-Cu deposits, N.W.T., Canada: Econ. Geology, *68*, 635-656.

Robinson, S.C., 1955, Mineralogy of uranium deposits, Goldfields, Saskatchewan: Canada Geol. Survey Bull. 31, 128 pp.

Ruzicka, V., 1971, Geological comparison between East European and Canadian uranium deposits: Canada Geol. Survey Paper 70-48, 196 pp.

Sarcia, J.A., Carrat, H., Poughon, A. and Sanselme, H., 1958, Geology of uranium vein deposits of France: Internat. Conf. Peaceful Uses of Atomic Energy, *2*, 592-611, United Nations.

Sassano, G.P., Fritz, P. and Morton, R.D., 1972, Paragenesis and isotopic composition of some gangue minerals from the uranium deposits of Eldorado, Saskatchewan: Canadian Jour. Earth Sci., *9*, 141-157.

Smith, E.E.N., 1974, Review of current concepts regarding vein deposits of uranium: in *Formation of Uranium Ore Deposits*, 515-529, Internat. Atomic Energy Agency, Vienna.

Szalay, A. and Samsoni, Z., 1969, Investigations on the leaching of uranium from crushed magmatic rocks: Geochem. International, *6*, 613-623.

Taylor, H.P., Jr., 1967, Oxygen isotope studies of hydrothermal mineral deposits: in Barnes, H.L., ed., *Geochemistry of Hydrothermal Ore Deposits*, 109-142, Holt, Rinehart and Winston, Inc., New York.

Tugarinov, A.I. and Naumov, V.B., 1969, Thermobaric conditions of formation of hydrothermal uranium deposits: Geochem. International, *6*, 89-103.

Viebig, W., 1905, Die Silber-Wismutgänge von Johanngeorgenstadt im Erzgebirge: Zeitschr. Prakt. Geologie, *13*, 89-115.

PART II

DESCRIPTIONS OF HYDROTHERMAL URANIUM DEPOSITS

I) NORTH AMERICA

A) CANADA

Figure 1 shows the locations of the major Canadian hydrothermal uranium deposits. Geological Survey of Canada Map 1252A, Mineral Deposits of Canada, indicates that all past and present economic uranium deposits in Canada, regardless of type, occur in, or in very close spatial association with, Aphebian (1.7-2.5 b.y.) or Paleohelikian (1.4-1.7 b.y.) rocks. The apparent absence in Canada of hydrothermal uranium deposits of Archean age (>2.5 b.y.) is striking.

1) SASKATCHEWAN

a) Beaverlodge District*

Location:

Uranium City, northwest Saskatchewan, Canada, 108° 25′ W, 59° 30′ N (Figures 1 and 2).

Geology:

See Geological Survey of Canada Map 1247A, which accompanies the report of Tremblay (1972), for the detailed geology of the district.

Figures 3 and 4 show the general geology of the Beaverlodge district and the locations of its more important mines and prospects. The Beaverlodge district is underlain by rocks of the Archean Tazin Group (basement complex) and Aphebian Martin Formation. Helikian gabbro dikes cut the rocks of both units. The Paleohelikian clastic sediments of the Athabasca basin crop out to the south of the district.

The Tazin Group consists of a thick sequence of interbedded graywackes, shales, sandstones, basic tuffs, and carbonates which have been regionally metamorphosed to amphibolite facies quartzites, argillites, amphibolites, impure marbles, garnet-biotite schist, and quartzo-feldspathic gneisses and schists. These rocks have been locally granitized to varying degrees; feldspathic quartzites appear to have been most readily and extensively granitized. Red granitic rocks occur commonly in the district. Tazin rocks in the Beaverlodge area are intensely folded and faulted, extensively brecciated and mylonitized, and are characterized by widespread hydrothermal alteration. All of the granites of the Beaverlodge area are thought to be of metasomatic origin, except for the granitic dikes and sills which are regarded as mobile fractions of the metasomatic granites. Tazin Group country rocks have slightly anomalous background uranium contents throughout the Beaverlodge district (see Table 1).

* Except as otherwise noted, this description is based on the authors' observations and the published accounts of Tremblay (1970, 1972), Little et al. (1972) and Robinson (1955).

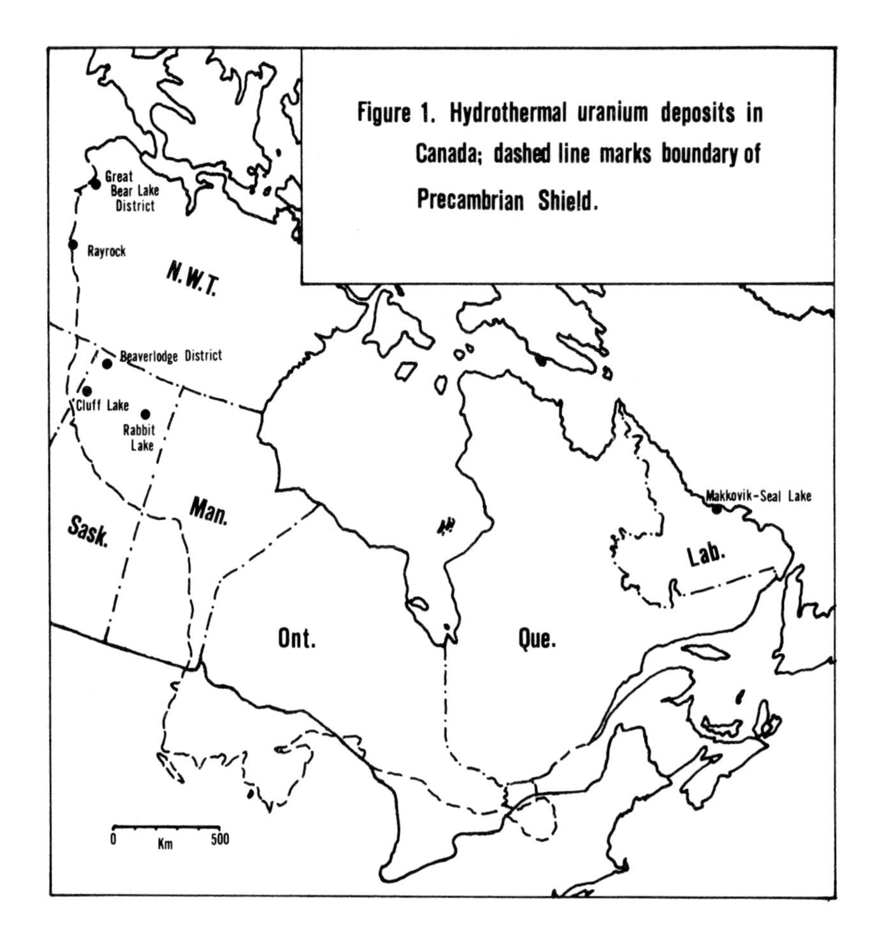

Figure 1. Hydrothermal uranium deposits in Canada; dashed line marks boundary of Precambrian Shield.

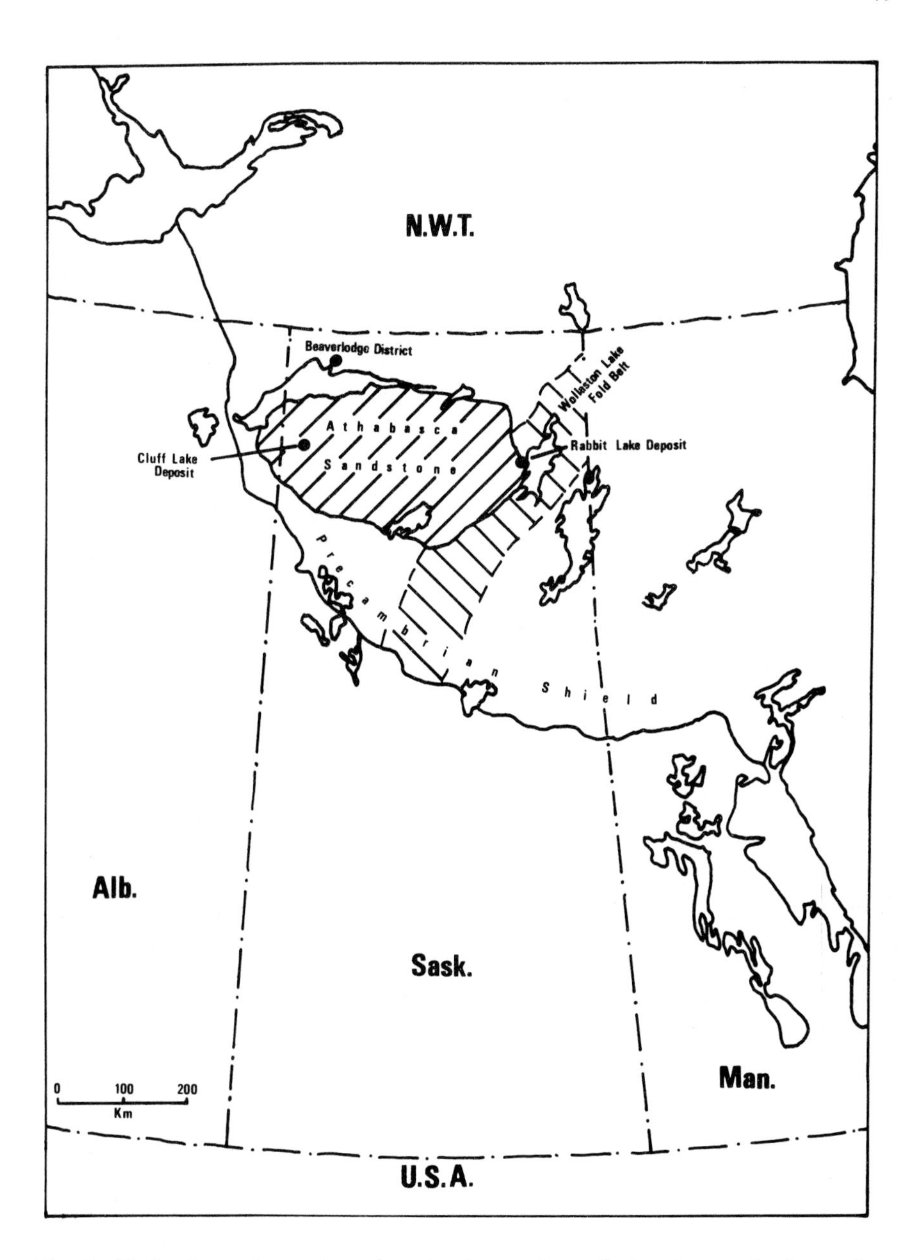

Fig. 2. Hydrothermal uranium deposits in northern Saskatchewan (redrawn from Knipping, 1974).

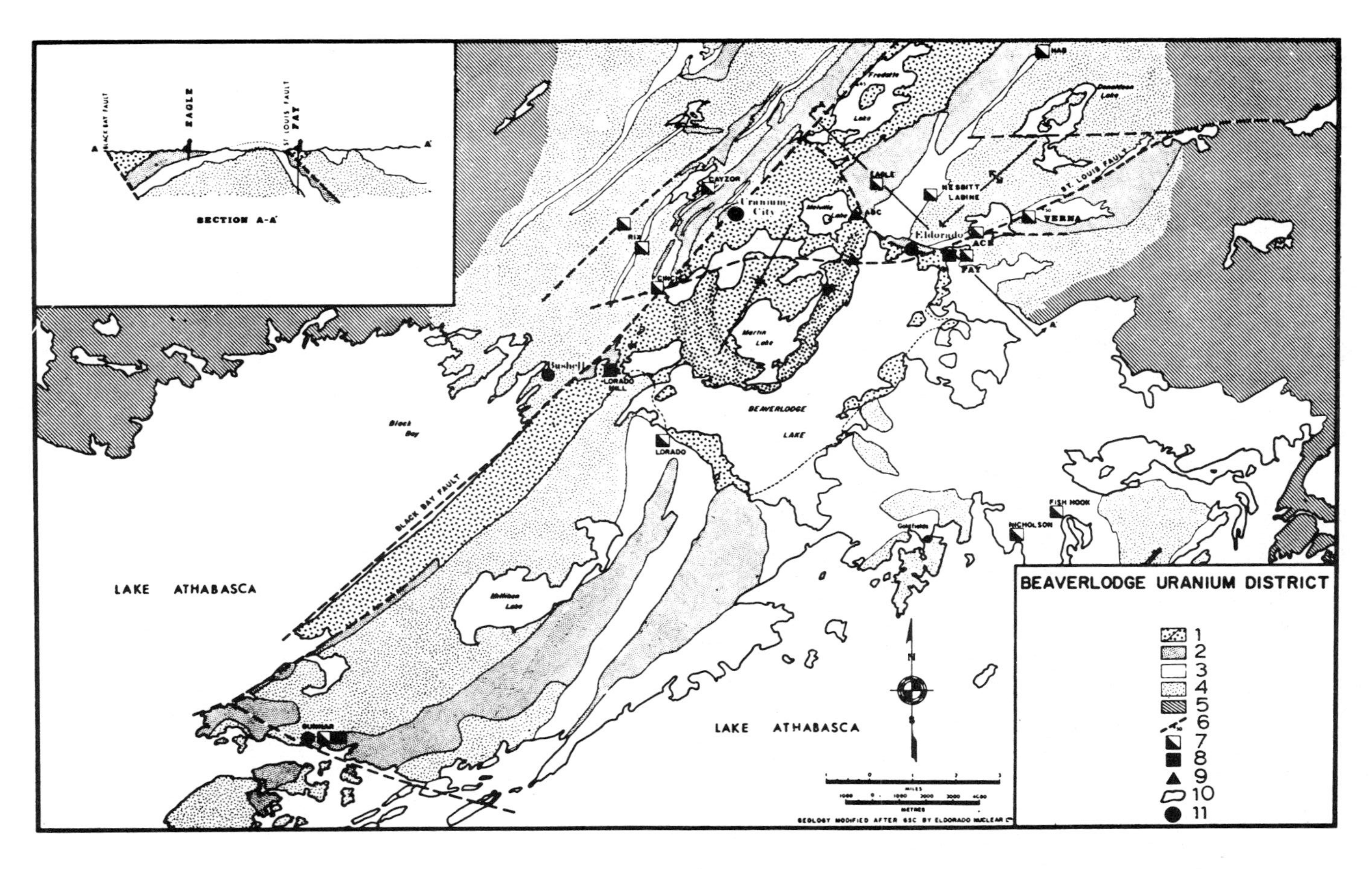

Fig. 3. Geology of the Beaverlodge district (from Little et al., 1972). 1 = Martin formation-seds./volc.; 2 = Upper Tazin red gneisses and mylonites; 3 = Tazin metasediments; 4 = Tazin granites and gneisses; 5 = undifferentiated Tazin; 6 = major fault;

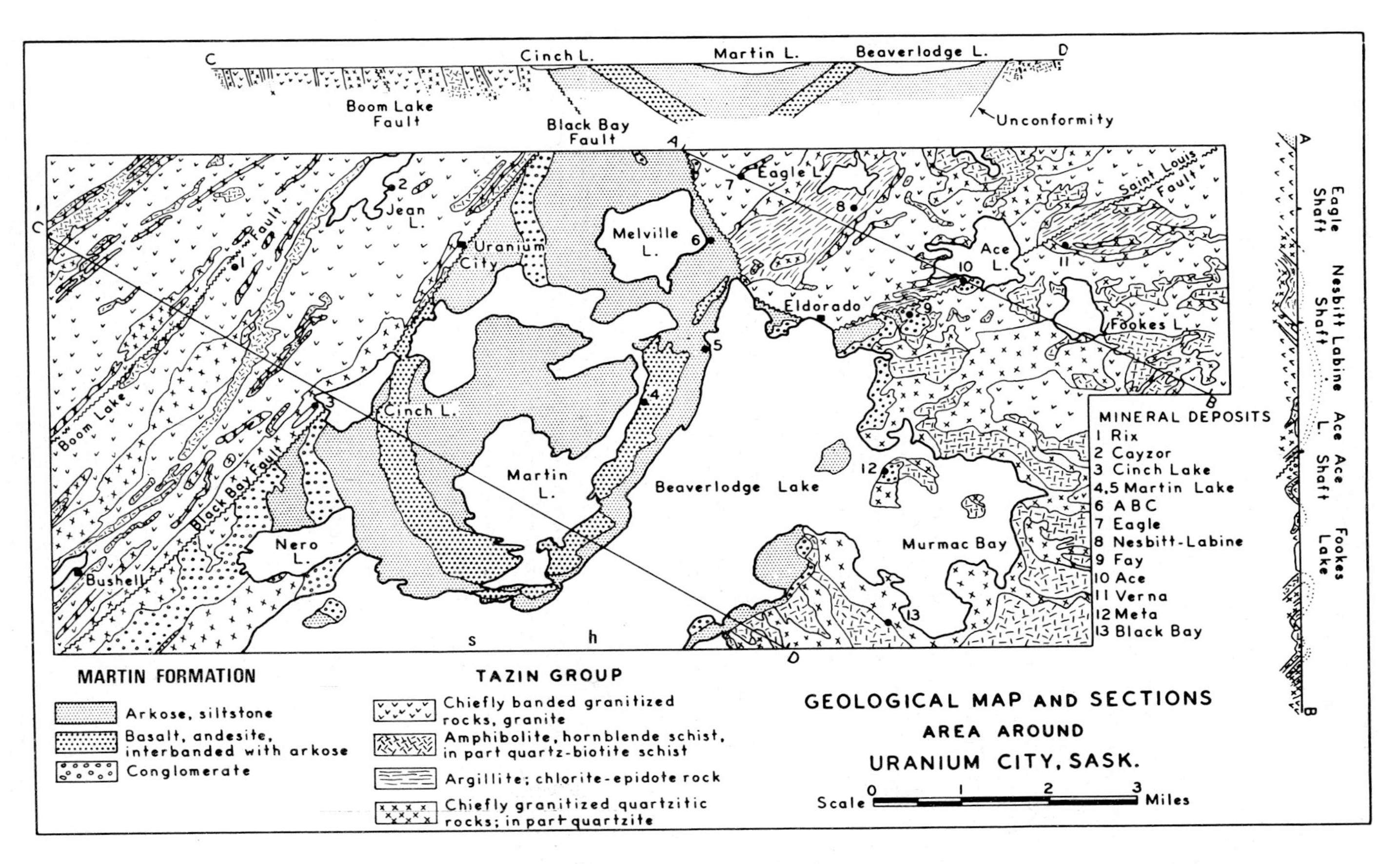

Fig. 4. Geology of the Beaverlodge district (modified from Tremblay, 1957).

Table 1. Uranium content of the main rock types in the Beaverlodge area (from Tremblay, 1972).

Rock types	Number of specimens analyzed		Uranium (ppm)	
Amphibolite	15			0.8 (0.3-2.7)
Argillite	6	14	4.6 (3.0-8.8)	4.5 (3.0-8.8)
Chlorite schist	3		4.4 (3.6-5.5)	
Impure quartzite	5		4.3 (3.4-6.8)	
White quartzite	7			3.0 (1.6-4.7)
Layered gneiss (white quartzite and chlorite schist)	6			3.7 (1.2-5.3)
Granite	15			5.4 (1.0-13.0)

Rocks of the Martin Formation unconformably overlie those of the Tazin Group. The Martin Formation consists of a red bed succession of polymict conglomerates, arkoses, siltstones, shales, andesitic volcanic flows and gabbroic sills. These rocks are unmetamorphosed and only gently folded. Several major post-Martin faults are present in the district; with respect to their spatial relation to the uranium deposits, the most notable of these are the east-northeast striking St. Louis and northeast trending Black Bay faults (Figure 4).

There exist two types of uranium occurrences in the Beaverlodge district:

(1) Syngenetic: monazite, uraninite, cyrtolite, uranothorite, pyrochlore-microlite and xenotime in pegmatites, granites and in certain islands of country rock in granitized terrane. No syngenetic uranium occurrences in the district are presently of economic interest.

(2) Epigenetic: Fine-grained pitchblende disseminations near and along fractures and pitchblende veins, fracture fillings, stockworks and mineralized breccias form locally commercial uranium deposits. Epigenetic uranium mineralization occurs in all rock types of the Beaverlodge district, but the major uranium deposits are restricted to altered and granitized portions of Tazin Group metasediments. Mica schist, epidotic amphibolite, and granitic gneisses are favorable Tazin host rocks. Beaverlodge uranium deposits show a close spatial relation to major faults, but ore usually occurs in subsidiary fault and fracture zones which have been brecciated and hematized. There appears to exist a direct relationship between the occurrence of uranium deposits and the intensity of granitization, alteration and mylonitization of Tazin Group host rocks.

Mineralogy:

See Table 2.

Pitchblende is the only important hypogene uranium mineral, but more than one generation of pitchblende is usually present in a deposit. This is shown by the occurrence of later pitchblende cementing brecciated earlier pitchblende as well as younger pitchblende veining older pitchblende. Pitchblende is commonly associated with hematite, chlorite and carbonate. Specular hematite occurs in some pitchblende veins.

Table 2. Mineralogy of the Beaverlodge District

Common	
albite	hisingerite
calcite	pitchblende
chlorite	pyrite
dolomite	quartz
fluorite	rammelsbergite
graphite	umangite
hematite	uraninite
Minor	
ankerite	magnetite
annabergite	malachite
arsenopyrite	niccolite
barite	nolanite
bornite	octahedrite
chalcopyrite	orthoclase
chalcomenite	pararammelsbergite
clausthalite	rutile
Cu-Co-Ni selenide	sericite
covellite	siderite
erythrite	siegenite
galena	sphalerite
garnet	tiemannite
gypsum	tourmaline
ilmenite	thucholite
klockmannite	uranophane
limonite	
Trace	
apatite	molybdenite
azurite	monazite
becquerelite	pyrrhotite
chalcocite	pyrochlore-microlite
cobaltite	serpentine
copper	silver
cuprosklodowskite	sklodowskite
dyscrasite	thorite
fergusonite	titanite
fourmarierite	ullmannite
gold	uranopilite
kasolite	uranothorite
liebigite	vandendriesscheite
marcasite	xenotime
masuyite	zippeite
meta-allanite	zircon (cyrtolite)

Paragenesis:

See Figures 5a and 5b.

Notes:
1) Where hematite, chlorite and calcite occur together in the wall rock, pitchblende is most likely to be present.
2) Pitchblende occurs where disseminated pyrite is abundant.
3) Quartz and pitchblende are not commonly associated; calcite and chlorite are the gangue minerals generally associated with the pitchblende.
4) The uranium mineralization at the Martin Lake mine is divisible into two periods of deposition; the first is characterized by the introduction of pitchblende, carbonate and hematite, and the second by the introduction of sulfides and selenides (Smith, 1952).
5) Marcasite is found where chlorite and carbonate have replaced pitchblende.

Zoning:

An increase of vanadium (occurring as nolanite) with depth has been reported.

Wall Rock Alteration:

Hematization is characteristic of the ore zones. It is often accompanied by the development of calcite, chlorite and red feldspar (adularia?) in the wall rocks (Christie, 1953). Other kinds of wall rock alteration recognized in the Beaverlodge deposits include chloritization (chlorite + sericite), epidotization (epidote + albite + chlorite + hematite + quartz), silicification, carbonatization and albitization. Titanium minerals, principally anatase and rutile, are found in both wall rocks and veins.

Age:

The following data are taken from Koeppel (1968). Syngenetic uranium minerals from basement pegmatites, etc. formed during two distinct periods, >2200 m.y. and 1930 ± 40 m.y. Initial epigenetic hydrothermal pitchblende mineralization occurred at about 1780 ± 15 m.y.b.p. This was followed by either three generations of pitchblende mineralization or three periods of lead loss. The first generation of hydrothermal pitchblende in the Beaverlodge district is the same age as the pitchblende of the Eldorado mine in the Great Bear Lake district, Northwest Territories. Tazin Group country rocks were metamorphosed and granitized 1930 ± 40 m.y. ago. The Martin Formation was deposited between 1780 ± 20 and 1930 ± 40 m.y. ago.

Ore Controls:

Major uranium deposits are located in intensely fractured, mylonitized and granitized Tazin Group rocks which have been hydro-

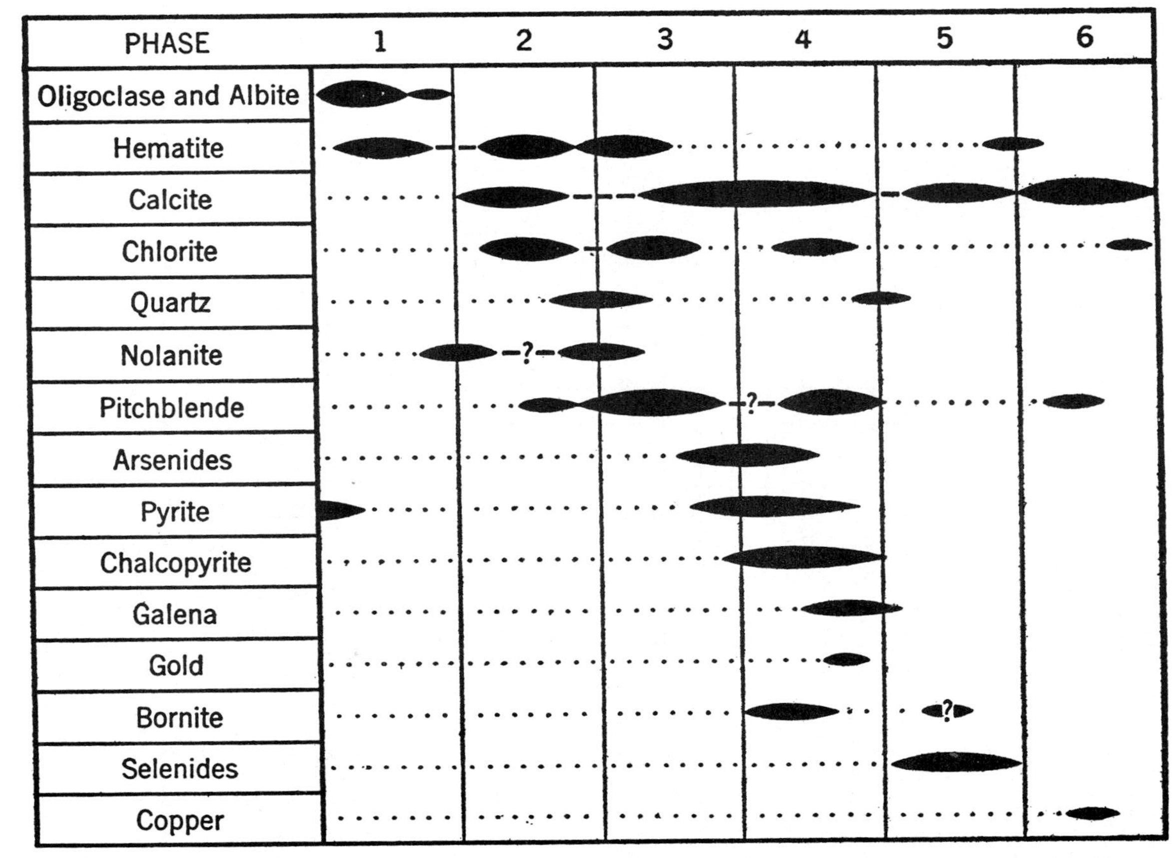

Fig. 5a. Paragenesis of the more common hydrothermal minerals of the Beaverlodge district (from Robinson, 1955).

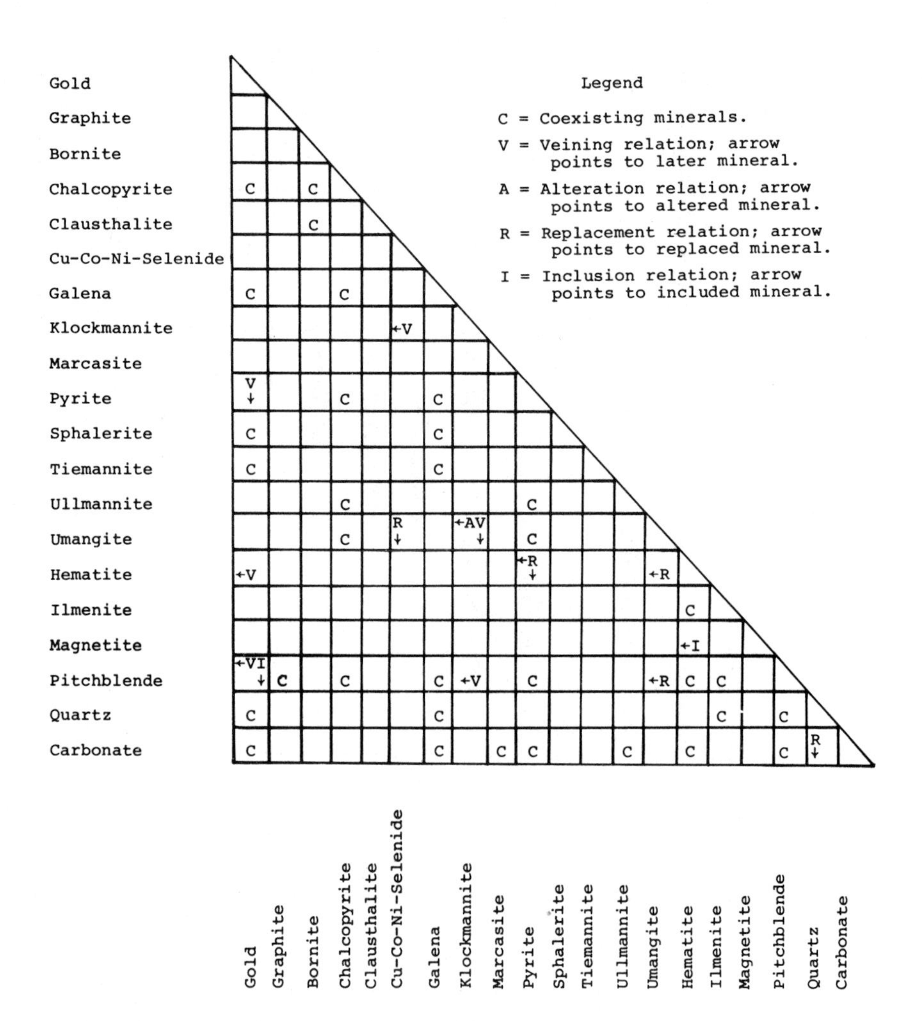

Fig. 5b. Paragenetic relations in the Beaverlodge District, Saskatchewan (modified from Petersen, 1973, unpublished report).

thermally hematized, chloritized, and locally carbonatized. In the eastern half of the map area of Tremblay (1972), there seems to be a close spatial relation between uranium deposits and Tremblay's Units 4a and 4b; these units consist of argillite, slate, chlorite-epidote rock, hornblende schist, and minor amphibolite.

(1) Fay-Ace-Verna Mine and Bolger Open Pit*

Location:

Beaverlodge district, north of Beaverlodge Lake along the St. Louis Fault (Figure 4).

Geology:

See Figure 4.

The Fay-Ace-Verna-Bolger deposit, the most productive of the district, consists of a complex network of pitchblende veins. Uranium ore bodies are distributed along the St. Louis fault over a length of more than 4 km. Ore bodies occur in subsidiary structures on both sides of the fault, but in general are not found in the main fault itself. Extensive mylonitization and brecciation characterize the St. Louis fault zone. North of the St. Louis fault the most important Tazin Group lithologies present in the mine area are highly epidotized and chloritized hornblende schist, fine-grained chloritic and sericitic argillite, and feldspathic quartzite. These rocks form part of the southeast limb of the Donaldson Lake anticline. South of the St. Louis fault large areas of granitized rocks occur. All rock types of the mine area have been feldspathized to some degree. Rock units in the vicinity of the ore bodies have been hematized; pitchblende-bearing structures generally have brick red walls. The thickness of alteration halos varies from centimeters to several meters depending on wall rock type. Mylonites and mica schists were most susceptible and epidotic amphibolites least susceptible to hematization. Ore extends from the surface to a depth greater than 1500 m.

The uranium deposits occurring in the footwall zone (north side) of the St. Louis fault are distributed over a distance of about 3 km east and west of the Ace shaft (Figures 4 and 6). All ore bodies are within 100 m of the fault and appear to be spatially related to a great bulge in the fault surface. Ore bodies of the footwall zone are of three types:

Breccia type: Uranium ore of this type is contained entirely within a brecciated and mylonitized feldspathic quartzite layer in contact with the St. Louis fault. Ore consists of angular rock fragments and dissemi-

* Description based on the authors' observations and the published accounts of Tremblay (1972) and Lang et al. (1962).

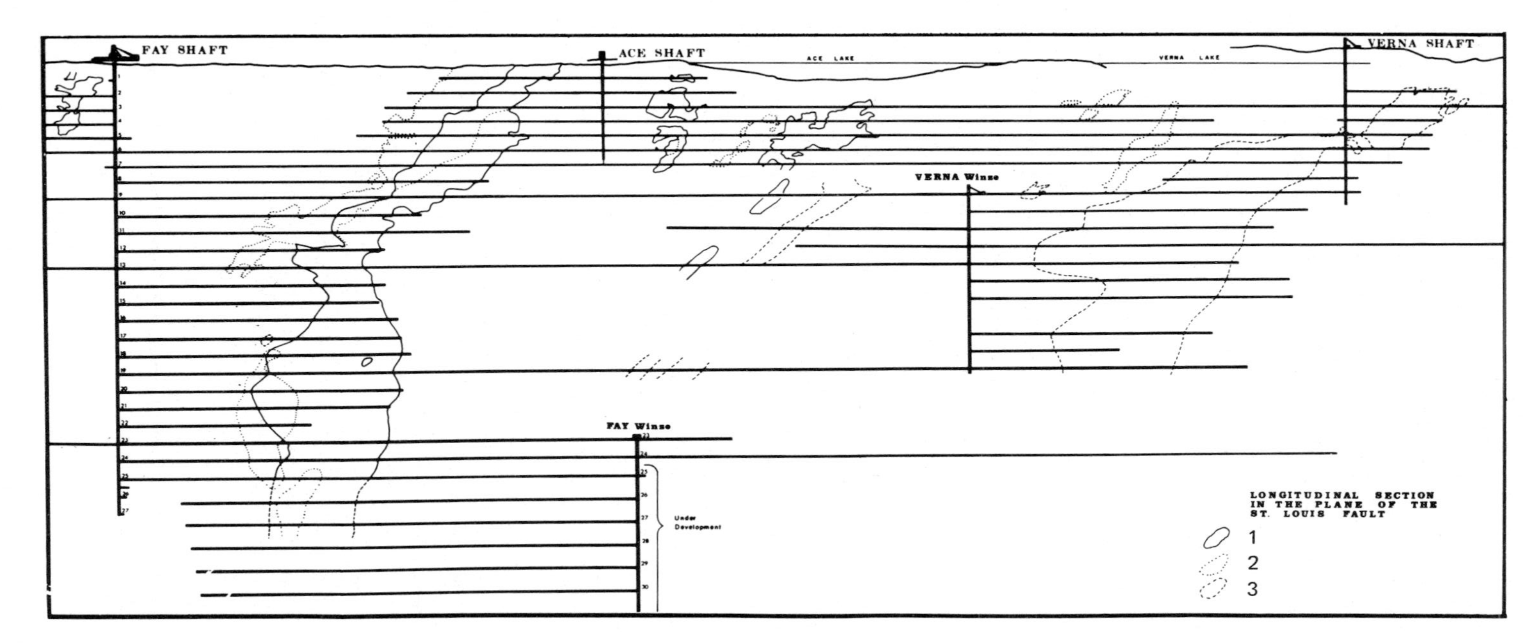

Fig. 6. Longitudinal projection of the ore bodies of the Fay—Ace—Verna mine (modified from Little et al., 1972). 1 = footwall orebody; 2 = immediate footwall orebody; 3 = hanging-wall orebody.

nated pitchblende in breccia matrix. Chlorite, calcite, and quartz are associated with pitchblende. Pyrite is also present and is locally abundant.

Stockworks type: Uranium ore of this type is contained entirely in feldspathic quartzite, and consists of a complex network of fine fractures filled with pitchblende, calcite, pyrite and chlorite.

Vein type: Uranium ore of this type occurs mainly along fractures which are sub-parallel to the St. Louis fault. Veins consist of pitchblende, calcite, chlorite, and locally fragments of wall rock. They contain in addition variable but generally small amounts of pyrite, clausthalite and nolanite. Trace amounts of chalcopyrite, galena, bornite, ilmenite, marcasite and sphalerite have also been reported.

Known ore bodies of the hanging wall zone (south side) of the St. Louis fault are distributed over a distance of more than 4 km, extending from east of the Verna shaft to west of the Fay shaft (Figures 4 and 6). The largest hanging wall uranium zones are the Verna ore bodies. These are all located at least 100 m from the main fault and occur in a wedge-shaped mass of argillite, hornblende schist and quartzite. Within this structural block, however, ore bodies are mostly confined to altered argillite. Hanging wall ore bodies are all of the vein and stockworks types; i.e., uranium mineralization is located along fractures. There is, however, some disseminated pitchblende in the wall rocks. In the fractures, pitchblende is associated with abundant carbonate and locally with much pyrite. A paragenetic diagram based on samples from the Fay mine and Bolger open pit is given in Figure 7. Sassano et al. (1972) have distinguished isotopically at least five generations of carbonate in Fay and Bolger samples.

(2) Gunnar Mine*

Location:

Beaverlodge district, Crackingstone Peninsula, north shore of Lake Athabasca, about 25 km southwest of Uranium City (Figure 3).

Geology:

See Figures 3 and 8.

The Gunnar deposit occurs on the east limb of a large anticline in a terrane composed of Archean Tazin Group metamorphosed and granitized sediments. Common rock types of the mine area are banded quartzofeldspathic paragneisses (with intercalated lenses of quartzite and mafic rocks), granites, and syenites. Some metamorphosed mafic

* Description based on the authors' observations and the published accounts of Williams et al. (1972), Canadian Mining Journal (1963) and Lang et al. (1962).

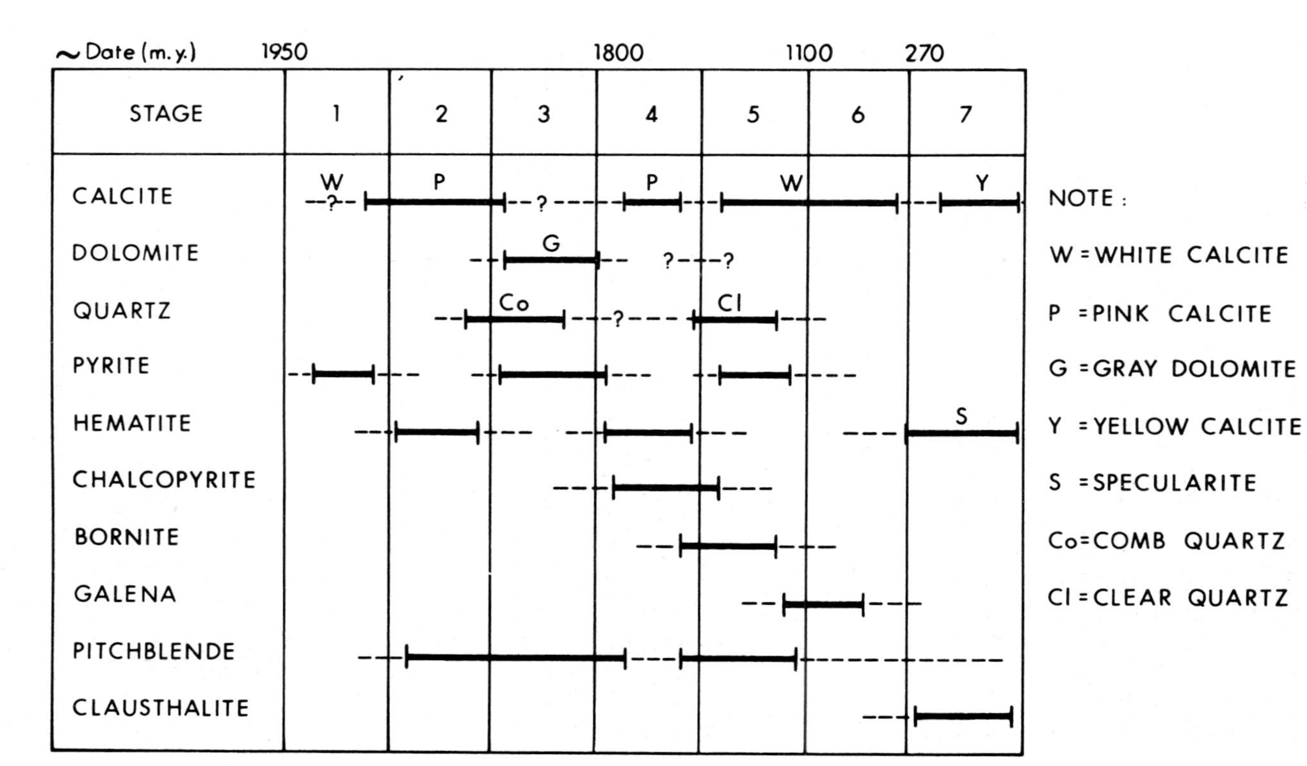

Fig. 7. Paragenesis of the more common hydrothermal minerals of the Fay mine and Bolger pit (modified from Sassano et al., 1972; reproduced by permission of the National Research Council of Canada from the Canadian Journal of Earth Sciences, v. 9, p. 144).

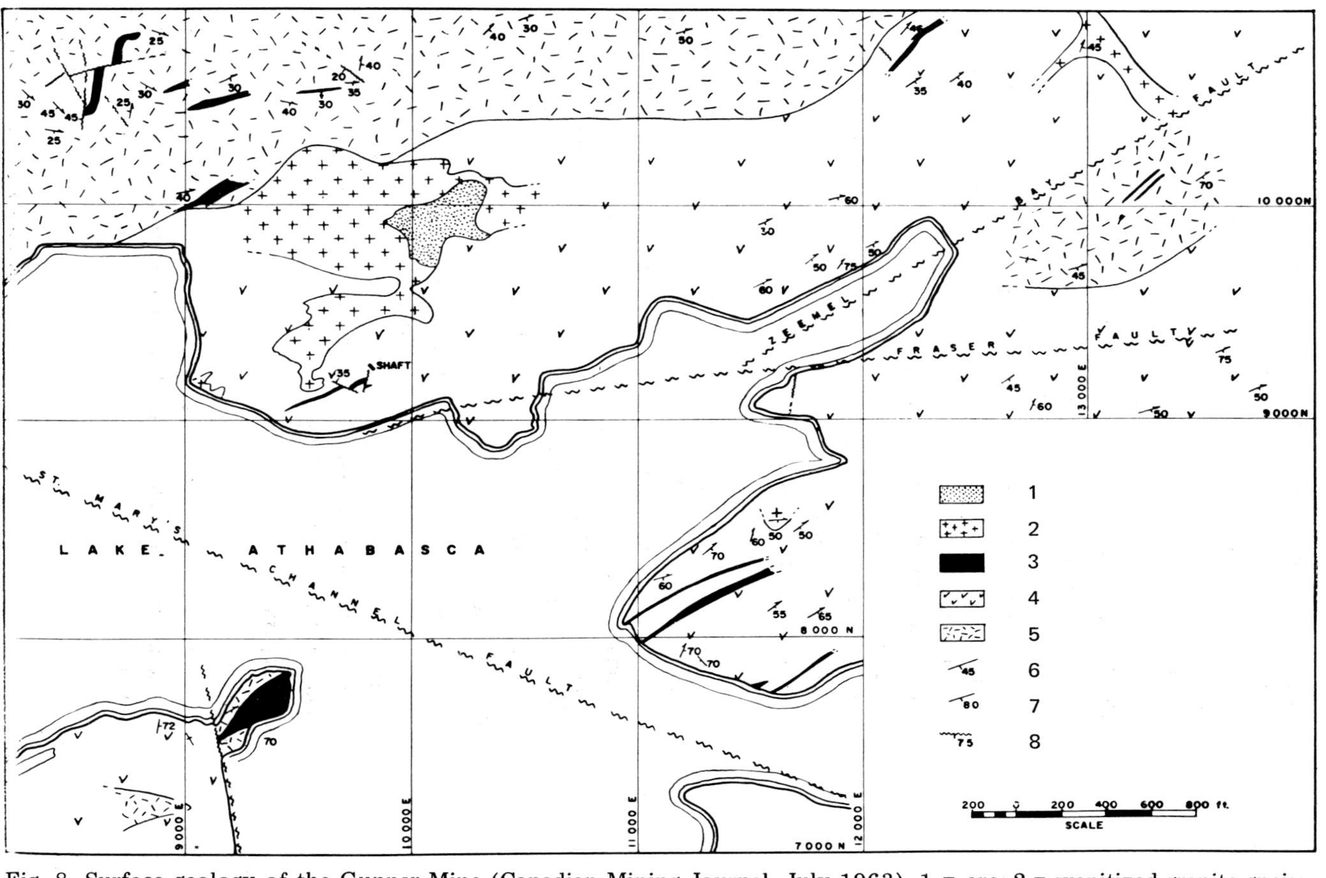

Fig. 8. Surface geology of the Gunnar Mine (Canadian Mining Journal, July 1963). 1 = ore; 2 = svenitized granite gneiss; 3 = mafics—chlorite and amphibole schist; 4 = granitic gneiss; 5 = paragneiss; 6 = bedding; 7 = foliation; 8 = fault.

units appear to cross-cut the felsic gneisses. The granites and syenites are products of the granitization of metasediments. Numerous large (½ m thick) veins consisting predominantly of epidote cross-cut granitic rocks near the mine.

The Gunnar deposit is located near the intersection of two major faults, the northeast striking Zeemel Creek and the east striking St. Mary's Channel fault. The pipe-like ore body consisted of stockworks and breccia zones in carbonatized syenitic rock. The syenite ore host rock, occurring as irregularly shaped bodies of various sizes within granitic gneiss, contains much albitic feldspar and carbonate, and only locally small amounts of secondary quartz. The syenite appears to have been derived from granitic gneiss through albitization, carbonatization, and desilication alteration processes, with carbonate replacing quartz. The granitic gneiss contains no carbonate, much quartz, and less fracturing and brecciation than the syenite. The syenite host rocks of the Gunnar ore appear to be similar to the uraniferous episyenites of the Limousin region of France.

Mineralogy:

Uranium ore minerals are pitchblende and minor uranophane; both are thought to be primary. No obvious supergene enrichment or impoverishment was noted at Gunnar. Most pitchblende was very finely disseminated, occurring as rims on albite grains in the syenite. Commonly pitchblende is closely associated with chlorite and hematite. Other introduced minerals present in Gunnar ore include calcite, dolomite, quartz, and traces of pyrite, chalcopyrite and galena.

Wall Rock Alteration:

Dark red hematitic alteration is common at Gunnar, but its occurrence is not always a guide to ore. Other alteration types present in the ore deposit are carbonatization, chloritization and albitization. Albitization preceded ore mineralization, whereas desilication and carbonatization took place during and after ore deposition.

(3) Lorado Mine*

Location:

Beaverlodge district, southwest end of Beaverlodge Lake on the Crackingstone Peninsula, about 8 km southwest of Uranium City (Figure 3).

* Description based on the authors' observations and the published account of Lang et al. (1962).

Geology:

See Figure 3.

The Lorado deposit occurs in Archean Tazin Group rocks on the east limb of a major northeast trending anticline whose axis extends along the Crackingstone Peninsula. The Tazin country rocks consist of quartzites, phyllites and schists. In addition, Aphebian conglomerates belonging to the Martin Formation crop out near the mine site.

Uranium ore occurs in a band of contorted graphitic and chloritic schists which are bounded on the east by quartzite and on the west by a pyritized shear zone. The Lorado deposit consists of several irregular pitchblende-bearing shoots. Most of the ore bodies are contained in a subsidiary syncline. The main ore control appears to have been structural (ore was localized in fractures along the axis and limbs of the subsidiary syncline), but the chemical control probably exerted by the reducing nature of the host rocks should not be overlooked. A notable feature of much of the Lorado ore is the absence of hematite and feldspar which occur abundantly in most of the other uranium deposits of the Beaverlodge district.

(4) Cayzor Mine*

Location:

Beaverlodge district, near Uranium City, west of the Black Bay fault (Figure 3).

Geology:

See Figure 4.

The Cayzor mine area is underlain by partly to completely granitized Archean Tazin Group metasediments, principally "meta-argillite" (chlorite-sericite schist) and quartzite. A dark green amphibolite and coarse-grained white to red granitic rocks are also present.

Individual ore bodies are generally small; they occur both in granitized chlorite schist and in impure quartzite. All ore occurs along fractures, but its distribution is erratic. The richest ore occurs at fault or fracture intersections. In a few cases, late basic dikes appear to have influenced the concentration of uranium ore along fault planes. Within mineralized zones pitchblende and thucholite, associated with pyrite, are distributed in pods, pockets and lenses along fractures. Carbonate is generally present with pitchblende, but hematitic alteration is rarely observed.

* Description based on Tremblay (1972) and Lang et al. (1962).

(5) Cinch Lake Mine*

Location:

Beaverlodge district, west of Black Bay fault, southwest of Uranium City (Figure 3).

Geology:

See Figure 4

The Cinch Lake mine is located near the Black Bay fault. North of the fault the area is underlain by Archean Tazin Group granitic and quartzitic layered gneisses. Martin Formation sediments and volcanics are found on the south side of the Black Bay fault. Ore bodies occur in a wedge shaped fault block of Tazin quartzo-feldspathic paragneiss and chloritic-feldspar augen gneiss which shows widespread cataclastic effects. The occurrence of ore is restricted to cataclastic zones. All rocks of the mine area exhibit some hematitic alteration, but near ore zones the rocks have an especially dark red color.

Individual ore shoots at Cinch Lake are generally small. Two types of uranium mineralization occur at the mine:

1) Disseminated type: pitchblende irregularly disseminated through mylonite, closely associated with calcite and chlorite; reddish brown, earthy hematite concentrated near pitchblende; pyrite, chalcopyrite, specular hematite and rutile occur locally.
2) Fracture filling and breccia type: mineralized tensional fractures and breccia zones occurring between two faults; breccia cemented and fractures filled by specular hematite, calcite, and pitchblende; intensified red hematitic alteration accompanies ore.

(6) Rix Mine*

Location:

Beaverlodge district, east of Boom Lake fault, west of Uranium City (Figure 4).

Geology:

See Figure 4.

The country rocks of the Rix mine area belong to the Tazin Group of Archean age and consist chiefly of paragneisses. Granitic phases grade into granitic banded gneisses. Post-Tazin faulting has produced a complex pattern of fracture, breccia, and mylonite zones. Brecciated and mylonitized rocks are the loci for most ore bodies at the Rix mine.

* Description based on Tremblay (1972) and Lang et al. (1962).

Calcite and chlorite are abundant in high grade ore zones.

At the Rix mine ore bodies are of two kinds:

1) Stockworks type: the most important type consists of stockworks of mineralized fractures with much disseminated pitchblende in adjacent wall rocks; mylonitized rocks are dark red due to hematization; uranium ore minerals are pitchblende and gummite.
2) Vein type: small, lenticular ore bodies of erratic distribution in the Boom Lake fault; pitchblende and carbonate fill tight fractures in siliceous units of an interlayered sequence of amphibolites and granitized siliceous rocks, which have been intruded by coarse-grained red granite dikes and sills; fracture fillings are composed of pitchblende mixed with carbonate, quartz, chlorite and wall rock fragments; small amounts of galena, chalcopyrite, hematite and pyrite are also present; disseminated pitchblende is found in wall rocks adjoining fractures.

b) Athabasca Basin

(1) Rabbit Lake Mine*

Location:

Northeast Saskatchewan, west of Wollaston Lake, 320 km southeast of Uranium City and 360 km north of La Ronge, 103° 43′ W, 58° 12′ N (Figure 2).

Geology:

See Figures 2 and 9.

The Rabbit Lake deposit is located in highly deformed and metamorphosed Aphebian rocks of the Wollaston Lake fold belt. These rocks are in fault contact with the unmetamorphosed and relatively undeformed Paleohelikian clastic sedimentary rocks of the Athabasca Formation which were deposited on an Aphebian erosion surface. Uranium mineralization is located in a synform structure consisting of Aphebian dolomitic and calcareous metasediments. These rocks have been regionally metamorphosed to cordierite-amphibolite grade during the Hudsonian orogeny (about 1.8 b.y.). In the axial zone of the synform structure the host rocks of the deposit are highly brecciated and altered. All known uranium ore is contained within the altered and brecciated zone of the metacarbonates.

*Description based on the authors' observations and the published accounts of Knipping (1974), Little (1974) and Little et al. (1972).

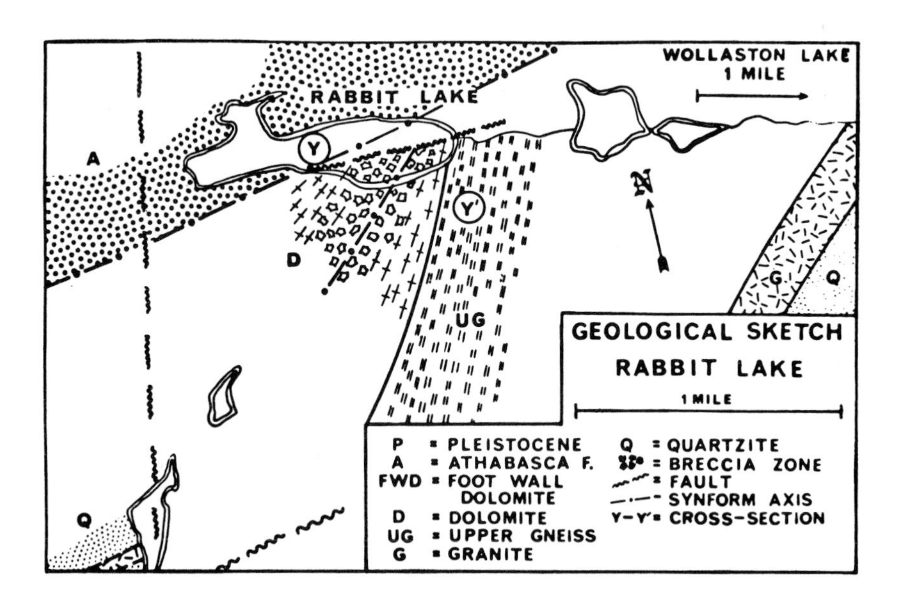

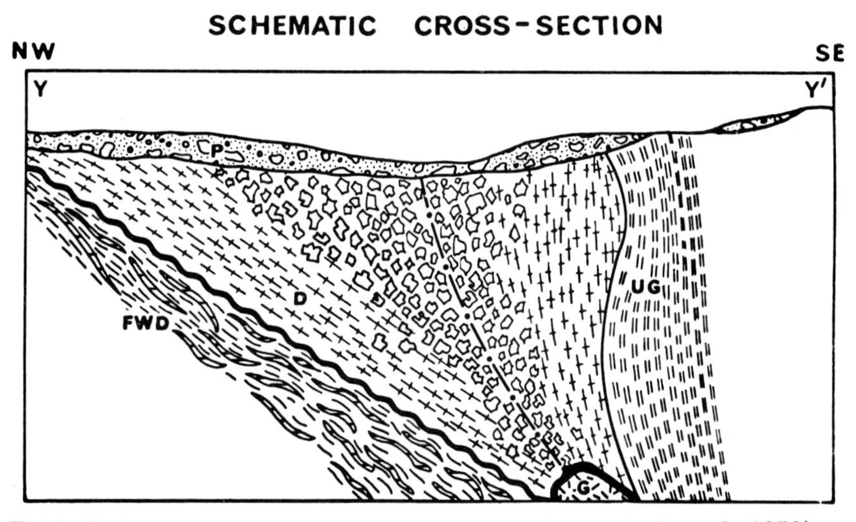

Fig. 9. Geology of the Rabbit Lake uranium deposit (from Little et al., 1972).

The ore-bearing sequence has been thrust over both Aphebian quartzites, dolomites, and calc-silicates and Paleohelikian Athabasca sandstone along a reverse fault. This fault limits the occurrence of uranium ore on the north side of the deposit; uranium mineralization has not been found in the fault, neither has it been found in the footwall rocks in spite of the fact that these rocks are strongly altered locally. Within the ore-bearing units of the hanging wall block, uranium mineralization appears to bottom out at a depth of about 100 m.

Pegmatites and granitic intrusions occur in the mineralized zone. These bodies were emplaced prior to the alteration event. To the south-east, the dolomitic ore host unit is in apparently conformable contact with an older, barren Aphebian unit, known as the "Upper Gneiss". This unit consists of well banded biotite paragneiss with intercalations of meta-arkose, dolomite, calc-silicate rock and quartzite. Pegmatite and granite sills and dikes are also common in this part of the Aphebian sequence. One pegmatite from the Rabbit Lake area contains 31 ppm uranium.

Uranium occurs chiefly as massive, colloform and sooty pitchblende, but coffinite is also present in small amounts. The pitchblende occurs as open space coatings; it coats breccia fragments, pore spaces and fracture walls. More than one generation of pitchblende deposition has been noted at Rabbit Lake.

Wall Rock Alteration:

Alteration of the host rocks at Rabbit Lake preceded the deposition of pitchblende. Host rocks have undergone widespread chloritization, sericitization, carbonatization and silicification, but argillization is largely confined to an envelope surrounding the ore zone. The rocks of the ore zone are characteristically green and soft. Alteration is very intense within 100 m of the present surface, but below this depth it weakens. At depths greater than 150 m, no alteration effects are known at Rabbit Lake. There is a striking paucity of hematization within the mineralized zone. Altered rocks at Rabbit Lake are rich in kaolinite, vermiculite, chlorite and dickite. In addition, small amounts of calcite, dolomite, quartz and hematite, and very small amounts of montmorillonite and magnesite are present. Various titanium minerals in the host rocks have been altered to anatase.

Mineralogy:

The primary uranium minerals are massive, colloform and sooty pitchblende and rare coffinite. The presence of thucholite has also been noted. Secondary uranium minerals present include uranophane, sklodowskite, boltwoodite and tyuyamunite; these are only found near the present erosion surface. Sulfides are present only in very small amounts. Pyrite, marcasite, galena and chalcopyrite are closely associated with pitchblende and may have been deposited at the same time. Pyrrhotite, sphalerite and niccolite are also present, but do not seem to be related to the pitchblende. Gangue minerals include calcite, quartz, dolomite and siderite.

Paragenesis:

1. Alteration of host rock.
2. Deposition of pitchblende.
3. Massive quartz and/or calcite.
4. Euhedral quartz.
5. Euhedral dolomite.
6. Sooty pitchblende.
7. Formation of secondary uranium minerals.

Ages:

The probable age of initial pitchblende deposition at Rabbit Lake is about 1.1 b.y. This age implies that uranium mineralization took place after, or near the end of, the deposition of the Athabasca sediments (1.1-1.7 b.y.) and that it is clearly not related to Hudsonian orogenic processes 1.8 b.y. ago.

(2) Cluff Lake Deposit*

Location:

Northwest Saskatchewan, Carswell Structure, 150 km south-southwest of Uranium City, 109° 32′ W, 58° 22′ N (Figure 2).

Geology:

See the Geological Map of Saskatchewan compiled by Whitaker and Pearson (1972).

The Cluff Lake uranium deposit consists of several ore bodies in, or associated with, deformed and metamorphosed basement rocks which form the central plug of the Carswell Structure. This structure is circular (35 km in diameter) and occurs as an island in the otherwise monotonous expanse of flat-lying clastic sedimentary rocks of the 100,000 km^2 Athabasca basin. The Carswell Structure consists of three concentric rock units:

(1) Central plug composed of Archean or Aphebian leucocratic metamorphic rocks and minor amphibolites.
(2) Crumpled ring of Paleohelikian Athabasca sandstone.
(3) Folded narrow band of Paleohelikian dolomite (Carswell For-

* Description based on Little et al. (1972) and von Backström (1974).

mation) which overlies the sandstone and forms the disconnected outer ring of the structure. The Carswell dolomite is not known to occur elsewhere in the Athabasca basin; it has been preserved in the Carswell structure because of slumping related to the formation of the structure.

Within the Carswell Structure both the Athabasca sediments and basement rocks are faulted and fractured, and basement rocks show retrograde metamorphic effects. Indications of shock metamorphism are present near the sandstone-basement contact.

At Cluff Lake, uranium mineralization is associated with anomalous concentrations of gold, vanadium, platinum, selenium, copper and nickel. Two types of ore bodies occur at the Cluff Lake deposit:

(1) Massive pitchblende-uraninite ore bodies in fractures cutting both sericitized and chloritized basement gneiss and Athabasca sandstone near their contact. For example, Pagel (1976) states that the important "D" ore body is located at the base of the Athabasca sequence in organic-rich pelitic material. In this deposit type crystalline uraninite and colloform pitchblende are associated with native gold, gold tellurides, clausthalite, copper sulfides and nickel-cobalt sulfarsenides (notably gersdorffite); the gangue is rich in chlorite.

(2) Disseminated pitchblende and coffinite mineralization in sheared phyllitic zones, containing chlorite and illite, and also in and near quartz veinlets; this type of deposit is found only in basement rocks; pitchblende and coffinite are associated with abundant sulfides of copper and lead; drilling has intersected crystalline uraninite at depth.

In both ore deposit types the host rocks are generally hematized, except in the immediate vicinity of the uranium deposits. Uraniferous asphaltites are present, and pitchblende has been observed to replace organic matter. Molybdenum is present, but not in a separate mineral phase.

Ages:

Uranium mineralization ages range from 1.1. to 1.4 b.y., which is probably contemporaneous with the deposition of part or all of the Athabasca Formation. On the other hand, Pagel (1976) reports an age of 1150 m.y. for the Cluff Lake deposit and states that the mineralization post-dates the Athabasca Formation. The Cluff Lake breccia (related to the formation of the Carswell Structure?) has been dated at 478 m.y.

2) NORTHWEST TERRITORIES

a) Great Bear Lake District*

Location:

East shore of Great Bear Lake, Northwest Territories, 118° W, 66° N. (Figure 1).

Geology:

See Figure 10 and Table 4.

Silver-uranium-copper-bismuth mineralization of the cobalt-nickel arsenide type is of widespread occurrence in the Echo Bay and Camsell River blocks of the Great Bear Lake district; however, uranium has been recovered from ore at only one locality, the Eldorado mine at Port Radium (Figure 10). The Echo Bay and Camsell River blocks are roof pendants of Late Aphebian volcanics and sedimentary rocks that occur within Hudsonian granites. These rocks have been little deformed, and have not been regionally metamorphosed above lower zeolite facies.

The Aphebian sedimentary and volcanic rocks of the Great Bear Lake district are divided into two major stratigraphic units, an older Echo Bay Group (composed of a thick sequence of intercalated tuffs, sediments and andesitic lava flows) and a younger Cameron Bay Group (composed of a thin sequence of poorly cemented, coarse-grained clastic sediments). Granitic rocks were intruded during and after the deposition of the rocks of the Echo Bay Group. The intrusion of large granite bodies tilted and formed contact metamorphic aureoles in the host sedimentary and volcanic rocks. Some replacement sulfide bodies were formed in intrusive contact zones. Sub-volcanic feldspar porphyries related to the granitic intrusions are petrographically indistinguishable from porphyritic lavas of the Echo Bay Group. The introduction of granitic rocks was followed by three periods of regional faulting, concomitant intrusion of diabase, and formation of giant quartz veins (see Furnival, 1935). The widely distributed giant quartz veins often contain uranium mineralization similar to that of the major deposits.

The earliest mineralization in the Great Bear Lake district is associated with acidic intrusive and volcanic rocks. It consists of sulfide (pyrite/marcasite, pyrrhotite, chalcopyrite, galena, and sphalerite) skarns, banded replacement zones, veins, and impregnations within the contact metamorphic aureoles of many of the granitic stocks. These occurrences, often referred to as fahlbands, are not related to the

*Except as otherwise noted, this description in based on Robinson and Badham (1974), Robinson and Ohmoto (1973) and Badham et al. (1972).

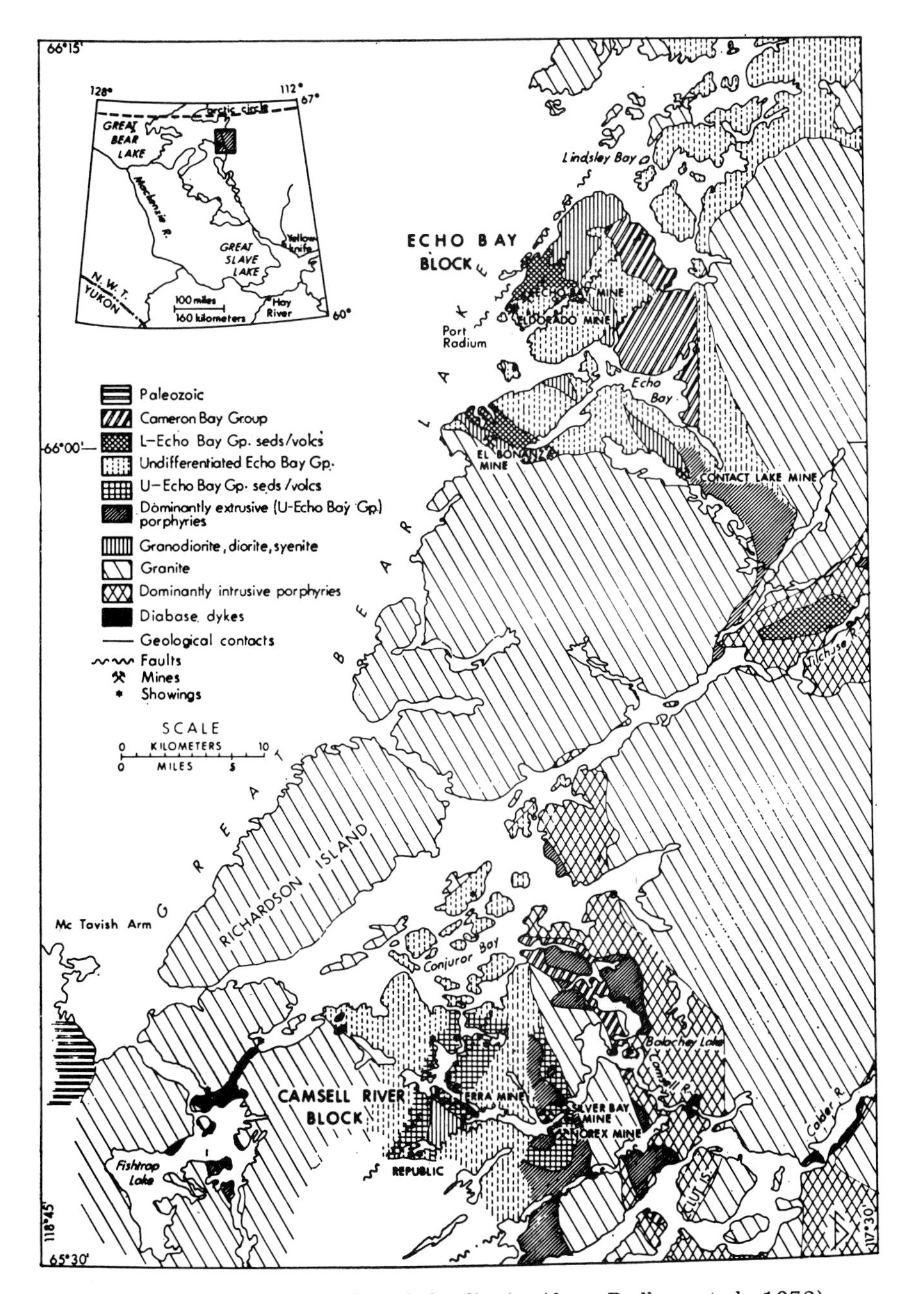

Fig. 10. Geology of the Great Bear Lake district (from Badham et al., 1972).

uraniferous vein mineralization. The later uranium-silver-cobalt-nickel-arsenic-bismuth-copper veins are localized by secondary faults, with most veins occurring in the red and green banded andesite tuffs of the lower part of the Echo Bay Group. The red tuff layers contain hematized feldspar. Locally, host tuffs contain as much as 10% fahlband-type pyrite. In all of the deposits of the district ore tends to be localized by the presence of concentrations of fahlband sulfides. In general, the mineralized veins of the district are short and thin, except at the Eldorado-Echo Bay deposit where vein thickness averages about 0.5 m.

Mineralogy and Paragenesis:

See Table 3 and Figures 11 and 12.

All vein deposits of the Great Bear Lake district have a similar paragenesis; large veins show the complete sequence, but small veins only contain parts of the general sequence.

Table 3. Mineralogy of the Great Bear Lake district

Common	
Chalcopyrite	Pyrite
Dolomite	Quartz
Galena	Rhodochrosite
Hematite	Silver
Pitchblende	
Minor	
Acanthite	Manganese oxides
Argentite	Mckinstryite
Bismuth	Safflorite-Rammelsbergite
Bornite	Skutterudite-Ni-Skutterudite
Chalcocite	Smaltite-Chloanthite
Cobaltite	Sphalerite
Covellite	Stromeyerite
Magnetite	Tetrahedrite
Trace	
Annabergite	Jalpaite
Arsenopyrite	Limonite
Barite	Löllingite
Chalmersite	Molybdenite
Erythrite	Niccolite
Fluorite	Polydymite
Freibergite	Pyrrhotite
Gersdorffite	Ruby Silver
Glaucodot	Witherite
Hessite	

Reported paragenetic relations:

1) Pitchblende is often accompanied by quartz, and cyclic pitchblende-quartz deposition has been frequently observed (Ruzicka, 1971). Campbell (1955) reported the interstitial deposition of quartz among pitchblende spherulites.

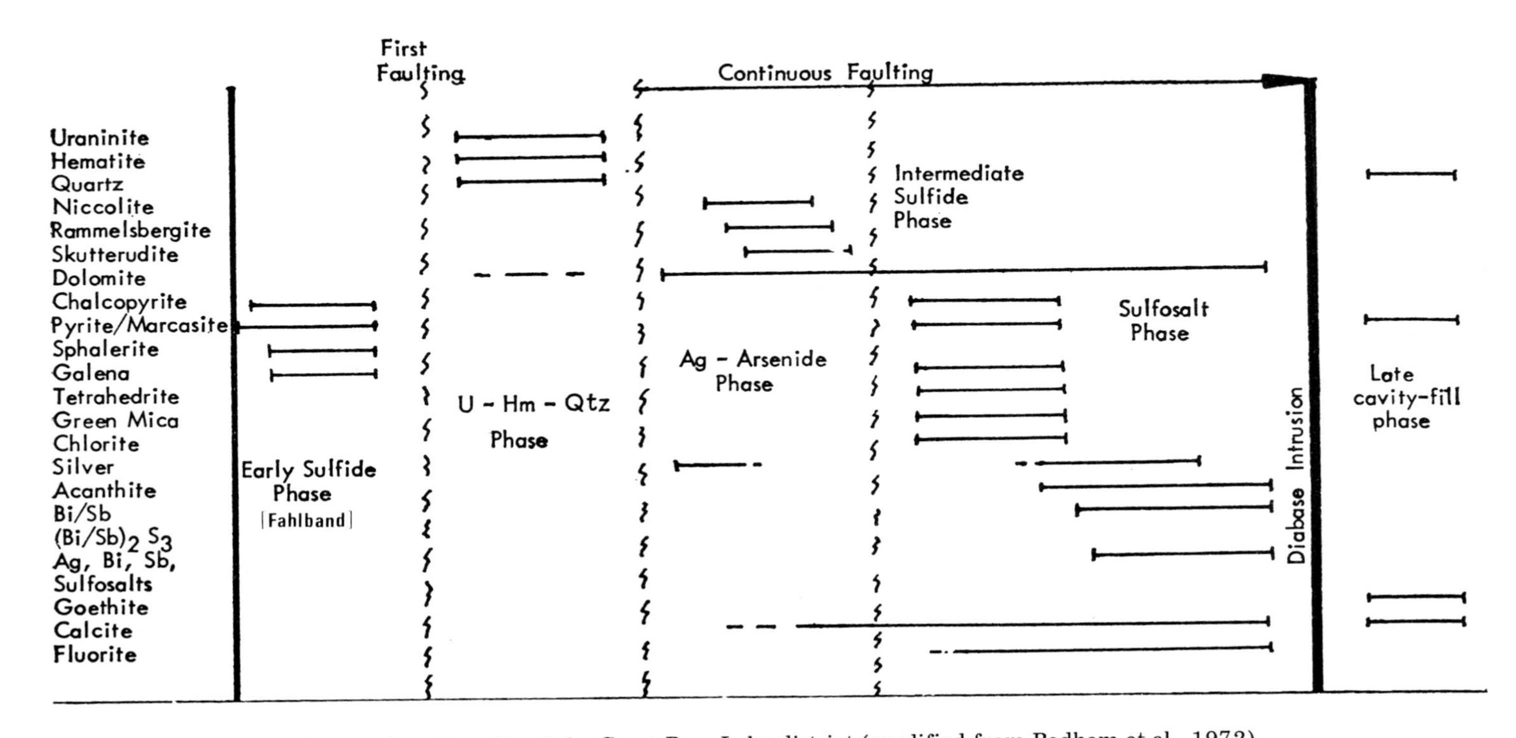

Fig. 11. Paragenesis of the uranium deposits of the Great Bear Lake district (modified from Badham et al., 1972).

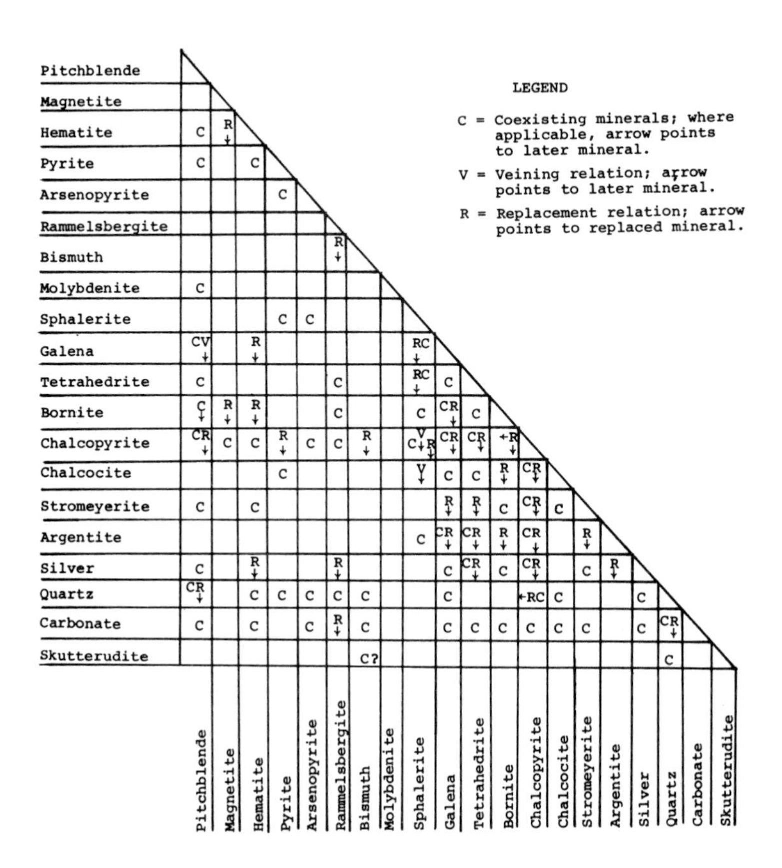

Fig. 12. Paragenetic relations in the Great Bear Lake District (modified from Petersen, 1973).

2) There is an apparent decrease in copper and an increase in silver during later stages of vein deposition.
3) Carbonate gangue is later than most of the quartz gangue.
4) The order of deposition of gangue is quartz, dolomite, chlorite, white mica, barite + siderite, and calcite + rhodochrosite (Jory, 1964).
5) Mckinstryite was the last primary sulfide deposited (Robinson and Ohmoto, 1973).
6) Arsenide grains are zoned, having nickel rich cores and cobalt-iron rich margins.

Wall Rock Alteration:

Veins are flanked by wall rock alteration halos of introduced microcline, hematite, chlorite, white mica and carbonate up to 3 m in thickness. The most intensive hematization occurs within 1.0-1.4 m of the veins.

Sequence of alteration after Donald (1956):

1) Early hematization associated with pitchblende deposition.
2) Argillization and sulfidization during deposition of vein quartz and cobalt-nickel minerals.
3) Chloritization during deposition of vein copper sulfides and chlorite.
4) Carbonatization during deposition of native metals and vein carbonate.

See Figure 14 for the wall rock alteration minerals at the Eldorado Mine.

Zoning:

Mineralization is nickel rich in the southeast and cobalt rich in the northwest part of the district. Nickel and cobalt mineralization has a broader distribution in the vein systems than pitchblende.

Ages:

Age relations for the Great Bear Lake district are summarized in Table 4.

Ore Controls:

Most ore is confined to the andesite tuff of the lower portion of the Echo Bay Group. Within the favorable rocks, ore is localized by faults and related fracture and breccia zones, and by concentrations of pre-vein fahlband sulfides.

Table 4. Stratigraphy and geochronology of the East Arm of Great Bear Lake (from Robinson and Ohmoto, 1973).

Stratigraphic Column	Geologic Events	Age (m.y.)
Paleozoics (?)		<600
	Uplift/Erosion	
	Intrusions of quartz-diabase sills & dikes/U-Ag mineralization/ Intrusions of giant quartz veins	~ 1,450
?		
	Faulting (NE trending)	
	Intrusions of diabase dikes	
	Faulting (E trending)	
Hornby Bay Group (300+m)	Subsidence	>1,500
	Uplift/Erosion	
	Regional metamorphism (?)	~ 1,650 (?)
?	Intrusion of granites & granodiorites	~ 1,800
	Intrusions of porphyries	
Cameron Bay Group (300+m)	Subsidence	~ 1,800
?	Uplift/Erosion	
Upper Echo Bay Sub-Group (1,500+m)	Volcanism	~ 1,800
Lower Echo Bay Sub-Group (1,200+m)		
Cliff Series (300m)	Volcanism	~ 1,800
Hornblende Porphyry Series (180 m)	Volcanism	~ 1,800
Tuff Series (390 m)	Volcanism	~ 1,800
Mine Series (350+m)	Volcanism	~ 1,800

(1) Eldorado/Echo Bay Deposit*

Location:

Great Bear Lake district, Port Radium, Northwest Territories (Figure 10). The Eldorado and Echo Bay mines occupy adjacent properties and mine different parts of the same ore deposit.

Geology:

The deposit is located within the Echo Bay structural block (see Figures 10 and 13a). Ore appears to be spatially related to the margins of bodies of porphyry (Figure 13b).

Mineralogy:

Major vein minerals are quartz, dolomite, hematite, native silver and chalcopyrite. In addition, lesser amounts of pitchblende, niccolite, rammelsbergite, skutterudite, calcite, marcasite, bornite, sphalerite, galena, acanthite, native bismuth and mckinstryite are present in the Eldorado/Echo Bay veins.

* Except as otherwise noted, this description is based on Robinson and Ohmoto (1973).

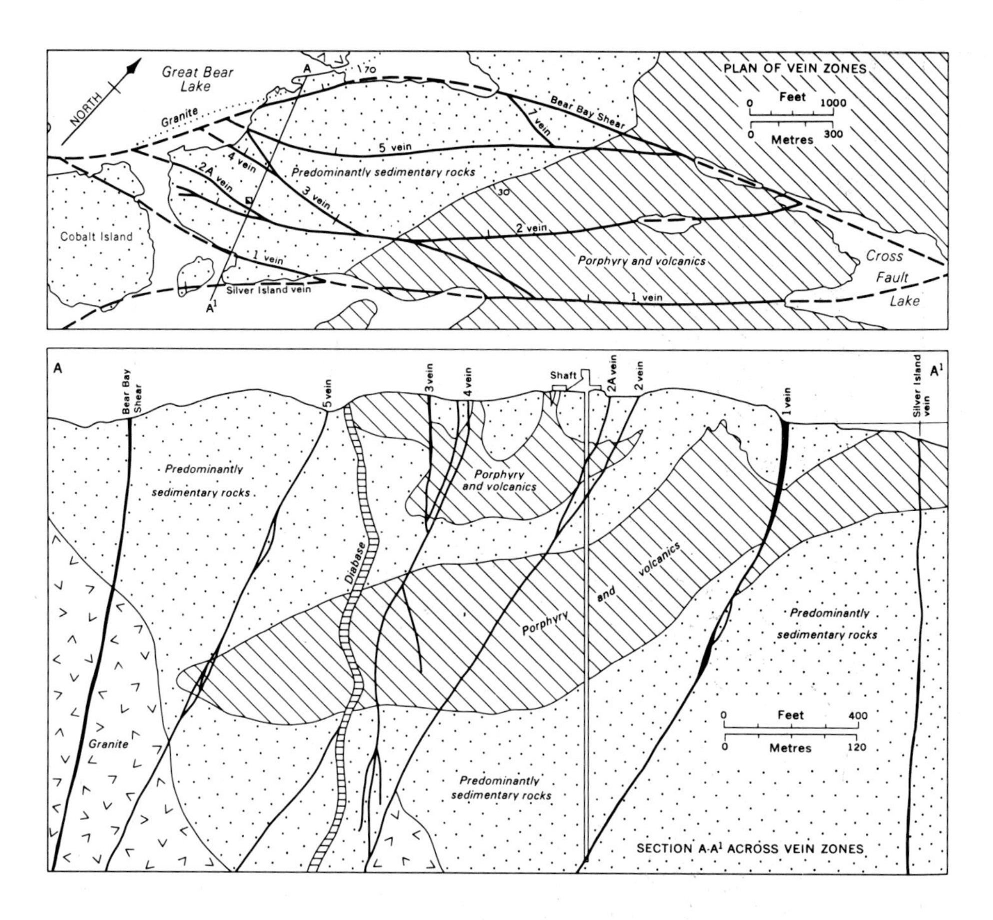

Fig. 13a. Generalized geology of the Eldorado mine (from Lang et al., 1970).

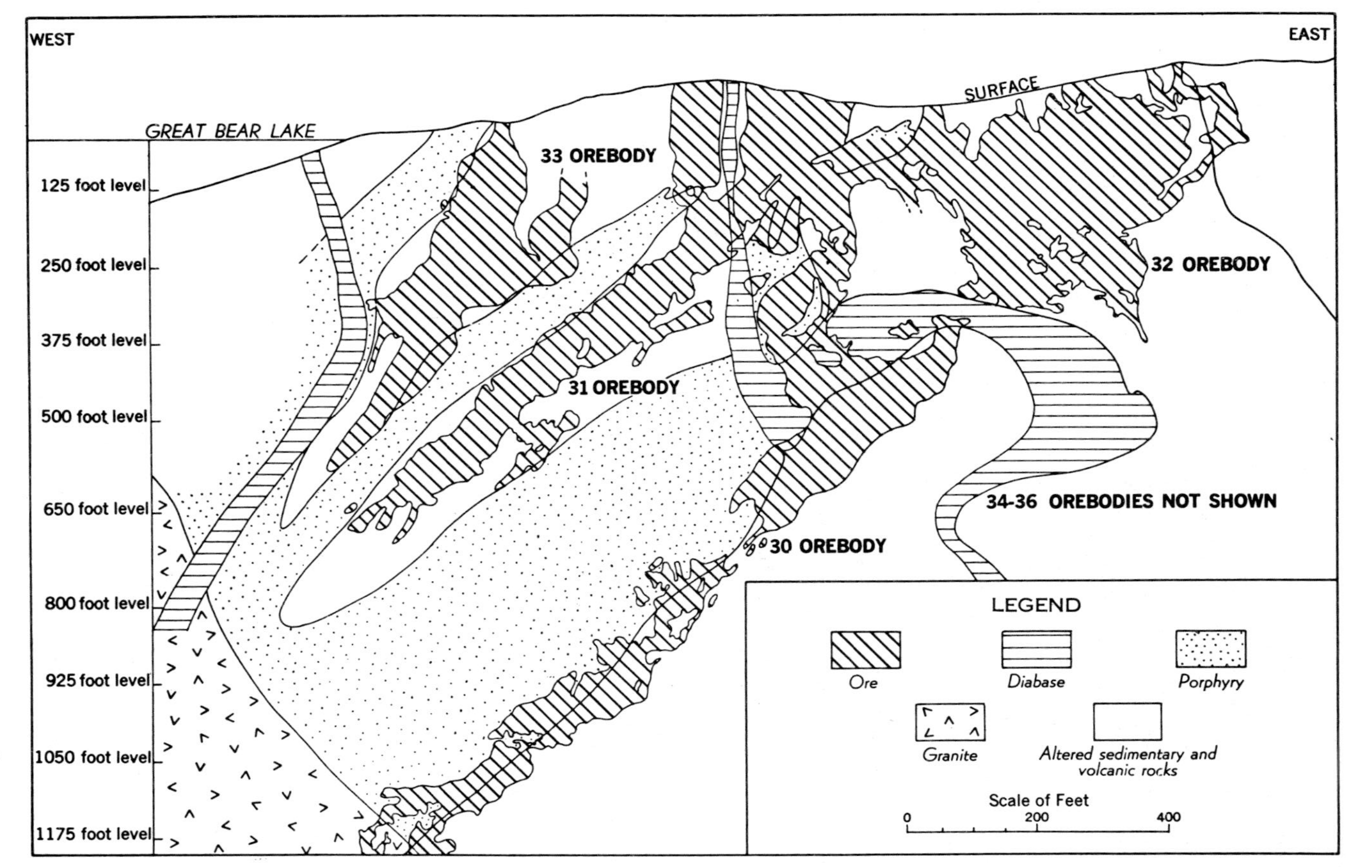

Fig. 13b. Longitudinal section of the No. 3 vein, Eldorado mine, showing the distribution of orebodies (from Lang et al., 1962).

Paragenesis:

See Figures 14 and 5-1 (Part I).

The generalized sequence of deposition is as follows:

1) Quartz + hematite.
2) Pitchblende.
3) Cobalt-nickel arsenides + dolomite + native silver.
4) Early sulfide stage (acanthite + marcasite).
5) Intermediate sulfide stage (acanthite + chalcopyrite + bornite + dolomite).
6) Late sulfide and native element stage (native silver + native bismuth, followed by mckinstryite).

(2) Terra Mine*

Location:

Great Bear Lake district, Camsell River block, 45 km south of Port Radium (See Figure 10).

Geology:

See Figure 10.

The Terra mine is the largest known silver-arsenide deposit in the Camsell River block. This deposit contains pitchblende mineralization, but uranium is not recovered from the silver ore. The paragenesis of the Terra deposit is presented in Figure 15 for comparison with those given for the Eldorado and Echo Bay mines. The generalized paragenetic sequence is as follows:

Pre-vein: Fahlband-type sulfide mineralization.

Stage 1: First phase of vein mineralization; characterized by the introduction of quartz, hematite, carbonate and pitchblende, accompanied by severe hematitization and silicification of wall rocks.

Stage 2: Deposition of native silver and arsenides.

Stage 3: Deposition of carbonates, base metal sulfides, and bismuth, silver and antimony as sulfosalts.

Stage 4: Deposition of late sulfides, carbonates and hematite in vugs.

Wall Rock Alteration:

The principal veins are surrounded by alteration halos containing introduced quartz, hematite, chlorite and carbonate.

* Description based on Badham (1975) and Robinson and Badham (1974).

Vein minerals	Stage 1	Stage 2	Stage 3	Stage 4	Stage 5	Stage 6
Apatite	—					
Quartz	—	—	— - —			
Hematite	—	—	-		—	—
Pitchblende		—				—
Ni-Co arsenides			—			
Pyrite				—		
Chlorite				—		
White mica				—		
Barite					—	
Siderite					—	
Dolomite			—		—	
Sphalerite					—	
Tetrahedrite					—	
Bornite				—	—	
Chalcopyrite				—	—	
Galena					—	
Calcite						—
Rhodochrosite						—
Silver minerals						—
Native bismuth						—
Wall-rock alteration minerals						
Apatite	—					
Microcline	—					
Hematite	—					
Quartz	— - —					
Chlorite				—		
White mica				—		
Pyrite				—		
Chalcopyrite				—	—	
Carbonates					— -	—

Fig. 14. Paragenesis of the Eldorado mine (from Ruzicka, 1971).

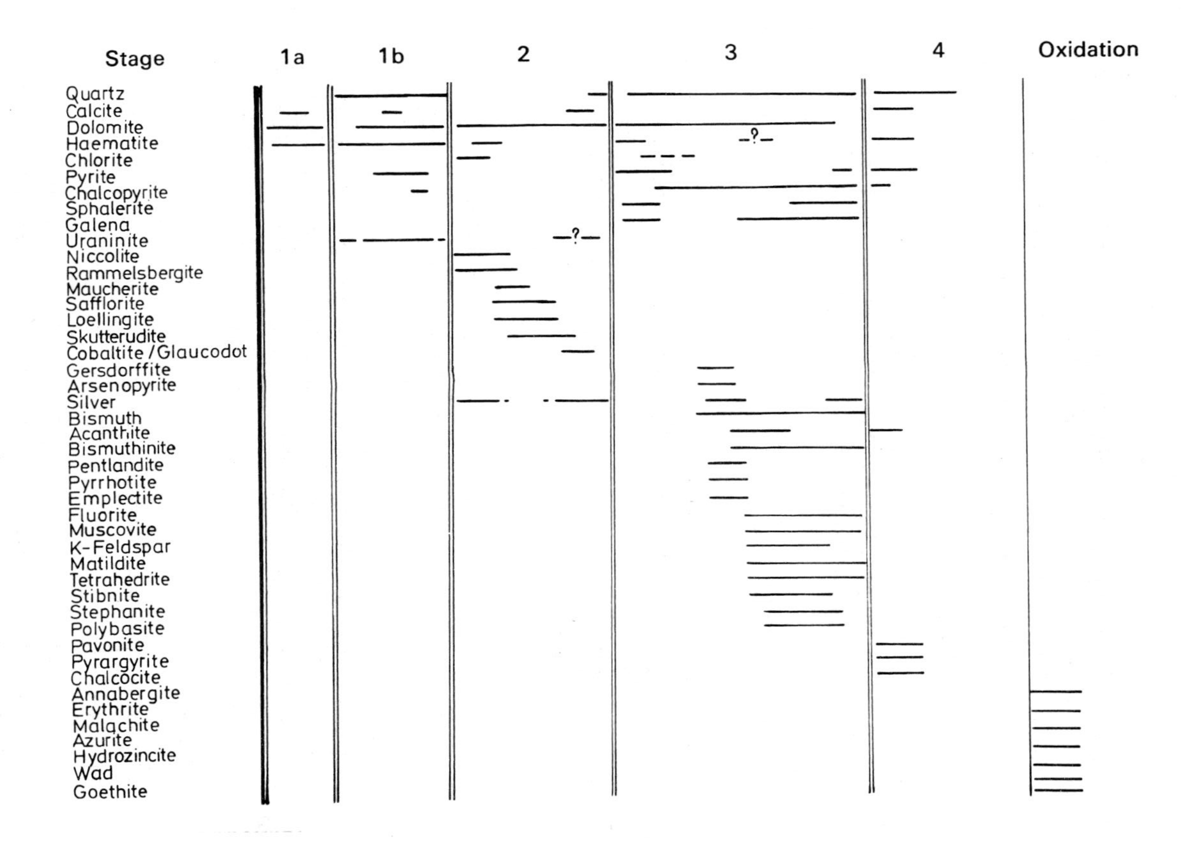

Fig. 15. Paragenesis of the Terra Mine vein mineralisation. Double lines represent main periods of dilation (from Badham, 1975).

Ore Controls:

Vein mineralization is most abundant in bedded volcanics; it is rarely found in intrusive rocks and is absent from volcanoclastic sediments. Mineralization is especially concentrated where country rocks contain abundant pre-vein (fahlband-type) sulfides.

b) Rayrock Mine*

Location:

Marian River region, Northwest Territories, 150 km northwest of Yellowknife, 63° 39′ N, 116° 30′ W (see Figure 1).

Geology:

See Figure 16.

The country rocks of the Rayrock mine area include quartzites, dolomites, argillites, cherts and mica schists of the Proterozoic Snare Group and the granites, granodiorites and quartz monzonites that intrude them. Uranium mineralization is associated with the prominent Marian River fault, and occurs in subsidiary fractures and breccia zones. The principal ore bodies at Rayrock are found in intensely fractured giant quartz veins and their altered wall rock envelopes (Figure 16). The quartz veins form stockworks within intrusive granodiorite. Ore is composed of fine-grained pitchblende, accompanied by small amounts of specular hematite, pyrite and chalcopyrite. Zones of altered granodiorite surround the quartz veins. These zones are characterized by epidotization and vary from 1.5 to 9 m in thickness; the thickest alteration aureoles are associated with strong uranium mineralization. Other wall rock alteration effects at Rayrock are hematization, silicification and chloritization. Hematization is well developed near radioactive zones.

3) LABRADOR

Makkovik—Seal Lake Uranium Deposits**

Location:

Central Labrador, 150 km east-northeast trending belt of uranium occurrences extending from Makkovik village on the Atlantic coast to Seal Lake inland, ~54° N, ~60° W (see Figure 1).

* Description based on Lang et al. (1962) and Ruzicka (1971).

** Description based on Greene (1974), Ruzicka (1971) and Gandhi et al. (1969).

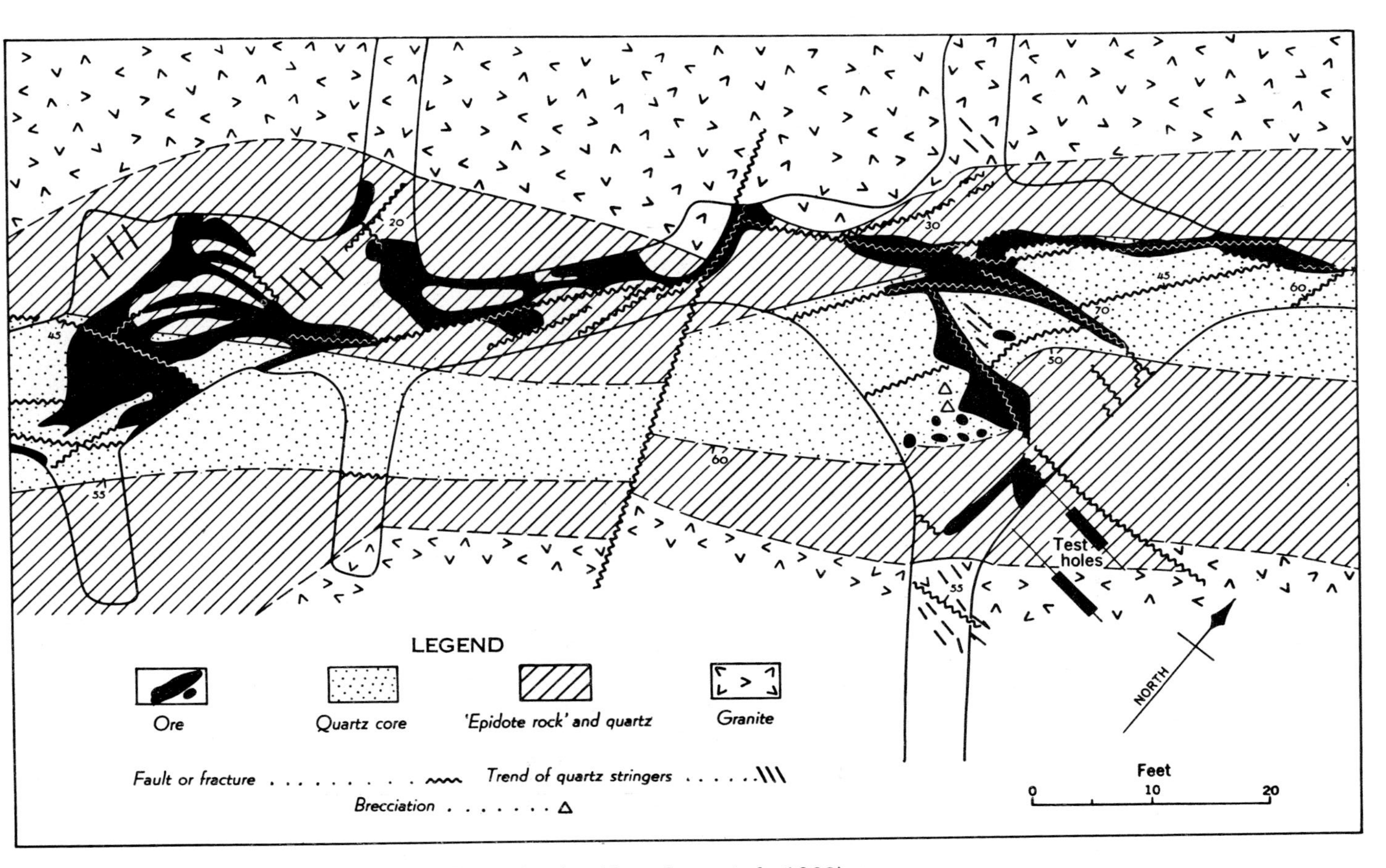

Fig. 16. Geology of part of the Adit level, Rayrock mine (from Lang et al., 1962).

Geology:

The country rocks of the Makkovik-Seal Lake belt consist of Archean basement (Hopedale banded gneisses), deformed Aphebian metasedimentary and metavolcanic rocks of the Aillik and Croteau Groups (feldspathic quartzite, conglomerate, paragneiss, and mafic lavas and tuffs), various Hudsonian intrusives (predominantly felsic), and late mafic intrusives. Both syngenetic and epigenetic uranium mineralization is known in the Makkovik-Seal Lake region. Syngenetic uranium occurs as uraninite in granites and pegmatites. Epigenetic uranium mineralization is found in several kinds of occurrences, all of which are hosted by Aphebian metasedimentary and metavolcanic rocks:

1) Pitchblende disseminations in feldspathic quartzites.
2) Pitchblende veins (quartz-carbonate) and disseminations in graphitic argillites, tuffs, and tuffites locally interbedded with amphibolites.
3) Pitchblende in shear and fault zones.
4) Pitchblende (botryoidal) mineralization in granulites, replacing hematite. Certain stratigraphic horizons seem to be favored by disseminated pitchblende mineralization.

Important Uranium Occurrences:

1) Kitts-Post Hill Zone: Uranium-carbonate-sulfide mineralization associated with a 22 km belt of mafic volcanics occurs as veins and disseminations in argillite, tuff, andesite and quartzite. The occurrence of the mineralization is controlled by the presence of tight folding. Minerals present include pitchblende, quartz, epidote, calcite, chalcopyrite, pyrite, and pyrrhotite.
2) Walker Lake-White Bear Mountain Area: Pitchblende occurs as disseminations and seams in porphyroblastic feldspathic quartzite.
3) Area South of Makkovik Village: Zones of red alteration and associated pitchblende seams occur along the contacts between mafic bands and quartzites. In addition, uranium mineralization occurs as pitchblende fracture fillings in mafic beds and as uraninite associated with magnetite and hematite in quartzites. Other minerals associated with the uranium mineralization include chalcopyrite, fluorite, sphalerite, pyrite and calcite.

Selected Canada References

Backström, J.W., von, 1974, Other uranium deposits: in *Formation of Uranium Ore Deposits,* 605-624, Internat. Atomic Energy Agency, Vienna.

Badham, J.P.N., 1975, Mineralogy, paragenesis and origin of the Ag-Ni, Co arsenide mineralization, Camsell River, N.W.T., Canada: Mineralium Deposita, *10*, 153-175.

Badham, J.P.N., Robinson, B.W. and Morton, R.D., 1972, The geology and genesis of the Great Bear Lake silver deposits: 24th Internat. Geol. Congress, Section 4, 541-548.

Barua, M.C., 1969, Geology of the uranium-molybdenum-bearing rocks of the Aillik-Makkovik Bay Area, Labrador: unpub. masters thesis, Queens University.

Campbell, D.D., 1955, Geology of the pitchblende deposits of Port Radium, Great Bear Lake, N.W.T.: unpub. doctoral thesis, Calif. Inst. Tech.

Canadian Mining Journal, 1963, The Gunnar story, Part II: Geology: *84*, no.7, 56-59.

Christie, A.M., 1953, Goldfields-Martin Lake map area: Canada Geol. Survey Mem. 269, 126 pp.

Dawson, K.R., 1956, Petrology and red coloration of wall-rocks, radioactive deposits, Goldfields region, Saskatchewan: Canada Geol. Survey Bull. 33, 46 pp.

Donald, K.G., 1956, Pitchblende at Port Radium: Canadian Mining Jour., *77*, no. 6, 77-79.

Fahrig, W.F., 1961, Geology of the Athabasca Formation: Canada Geol. Survey Bull. 68, 41 pp.

Furnival, G.M., 1935, The large quartz veins of Great Bear Lake, Canada: Econ. Geology, *30*, 843-859.

Gandhi, S.S., Grasty, R.L. and Grieve, R.A.F., 1969, The geology and geochronology of the Makkovik Bay area, Labrador: Canadian Jour. Earth Sci., *6*, 1019-1035.

Greene, B.A., 1974, An outline of the geology of Labrador: Newfoundland Dept. of Mines and Energy Inf. Circ. 15, 64 pp.

Jory, L.T., 1964, Mineralogical and isotopic relations in the Port Radium pitchblende deposit, Great Bear Lake, Canada: unpub. doctoral thesis, Calif. Inst. Tech.

Knipping, H.D., 1974, The concepts of supergene versus hypogene emplacement of uranium at Rabbit Lake, Saskatchewan, Canada: in *Formation of Uranium Ore Deposits,* 531-549, Internat. Atomic Energy Agency, Vienna.

Koeppel, V., 1968, Age and history of the uranium mineralization of the Beaverlodge area, Saskatchewan: Canada Geol. Survey Paper 67-31, 111 pp.

Lang, A.H., Griffith, J.W. and Steacy, H.R., 1962, Canadian deposits of uranium and thorium: Canada Geol. Survey Econ. Geology Series No. 16, 2nd ed., 324 pp.

Lang, A.H., Goodwin, A.M., Mulligan, R., Whitmore, D.R.E., Gross, G.A., Boyle, R.W., Johnston, A.G., Chamberlain, J.A. and Rose, E.R., 1970, Economic minerals of the Canadian Shield: in Douglas, R.J.W., ed., *Geology and Economic Minerals of Canada,* Canada Geol. Survey Econ. Geology Rept. No. 1, 151-226.

Little, H.W., 1970, Distribution of types of uranium deposits and favorable environments for uranium exploration: in *Uranium Exploration Geology, 35-48,* Internat. Atomic Energy Agency, Vienna.

Little, H.W., Smith, E.E.N. and Barnes, F.Q., 1972, Uranium deposits of Canada: 24th

Morton, R.D., 1974, Sandstone-type uranium deposits in the Proterozoic strata of northwestern Canada: in *Formation of Uranium Deposits, 255-273, Internat. Atomic Energy Agency, Vienna.*

Morton, R.D. and Sassano, G.P., 1972, Structural studies on the uranium deposit of the Fay Mine, Eldorado, northwest Saskatchewan: Canadian Jour. Earth Sci., *9,* 803-823.

Pagel, M., 1976, La diagenèse des grés et les gisements d'uranium du bassin Athabasca (Canada): Rapport Annuel 1975, Centre de Recherches Pétrographiques et Géochimiques, 50-52.

Petersen, U., 1973, Hydrothermal uranium deposits: unpublished report for Exxon Corporation.

Robinson, B.W. and Badham, J.P.N., 1974, Stable isotope geochemistry and the origin of the Great Bear Lake silver deposits, N.W.T., Canada: Canadian Jour. Earth Sci. *11,* 698-711.

Robinson, B.W. and Morton, R.D., 1972, The geology and geochronology of the Echo Bay area, Northwest Territories, Canada: Canadian Jour. Earth Sci., *9*, 158-171.

Robinson, B.W. and Ohmoto, H., 1973, Mineralogy, fluid inclusions, and stable isotopes of the Echo Bay U-Ni-Ag-Cu deposits, N.W.T., Canada: Econ. Geology, *68*, 635-656.

Robinson, S.C., 1955, Mineralogy of uranium deposits, Goldfields, Saskatchewan: Canada Geol. Survey Bull. 31, 128 pp.

Ruzicka, V., 1971, Geological comparison between East European and Canadian uranium deposits: Canada Geol. Survey Paper 70-48, 196 pp.

Sassano, G.P., Fritz, P. and Morton, R.D., 1972, Paragenesis and isotopic composition of some gangue minerals from the uranium deposits of Eldorado, Saskatchewan: Canadian Jour. Earth Sci., *9*, 141-157.

Shegelski, R.J., 1973, Geology and mineralogy of the Terra silver mine, Camsell River, Northwest Territories: unpub. masters thesis, Toronto.

Smith, E.E.N., 1952, Structure, wall-rock alteration and ore deposits at Martin Lake: unpub. doctoral thesis, Harvard, 125 pp.

Tremblay, L.P., 1957, Ore deposits around Uranium City: in *Structural Geology of Canadian Ore Deposits, 2*, 211-220, Canadian Inst. Mining and Metallurgy, Montreal.

Tremblay, L.P., 1970, The significance of uranium in quartzite in the Beaverlodge area, Saskatchewan: Canadian Jour. Earth Sci., *7*, 280-305.

Tremblay, L.P., 1972, Geology of the Beaverlodge mining area, Saskatchewan (revised): Canada Geol. Survey Mem. 367, 265pp.

Watkinson, D.H., Heslop, J.B. and Ewert, W.D., 1975, Nickel sulfide-arsenide assemblages associated with uranium mineralization, Zimmer Lake area, northern Saskatchewan: Canadian Mineralogist, *13*, 198-204.

Whitaker, S.H. and Pearson, D.E., 1972, Geological map of Saskatchewan (1:1,267,200): Saskatchewan Research Council, Saskatoon.

Williams, R.M., Little, H.W., Gow, W.A. and Berry, R.M., 1972, Uranium and thorium in Canada: 4th Internat. Conf. Peaceful Uses of Atomic Energy, *8*, 37-57, United Nations.

B) UNITED STATES

Hydrothermal uranium deposits presently account for less than 4% of the total U.S. production plus known reserves of uranium. Although there exist several hundred hydrothermal uranium occurrences in the United States, only a few large deposits of this type are known. Fewer than 20 occurrences have been developed into important mines, and only 6 of these have production plus reserves in excess of 500 tons U_3O_8 (Figure 17): the Schwartzwalder mine in the Colorado Front Range, the Pitch mine in the Marshall Pass area, Colorado, the Los Ochos mine in the Cochetopa district, Colorado, the Marysvale district, Utah, the Orphan mine in the Grand Canyon, Arizona, and the Midnite mine near Spokane, Washington. These uranium deposits, and several others of lesser importance, are described in the following pages. The locations of several hundred hydrothermal uranium deposits in the United States are shown on Plate I of Walker et al. (1963).

1) ARIZONA

Orphan Mine*

Location:

South rim of Grand Canyon, 3 km west of Grand Canyon Village, 112° 7′ W, 36° 3′ N (Figure 17).

Geology:

Uranium mineralization at the Orphan mine occurs in and about the margins of a nearly circular, vertical breccia pipe which crops out at the contact between the Coconino Sandstone and the Hermit Shale on the south wall of the Grand Canyon. The pipe extends more than 400 m downward from the lower part of the Coconino Sandstone, through the Hermit Shale, into the Supai Formation, and may continue into the Redwall Limestone beneath. There is no evidence to suggest that sedimentary units above the Coconino were involved in the formation of the pipe. The breccia pipe probably formed as a result of solution collapse in the Redwall Limestone. The evidence suggests subsidence of at least 200 m in the center of the pipe structure. Host rocks for the breccia pipe are unmetamorphosed and only mildly deformed Upper Paleozoic sedimentary units of the Colorado Plateau. Miocene to Late Pleistocene lava flows are found at several localities in the Grand Canyon area, but no occurrences of post-Precambrian intrusive rocks are known.

The pipe boundary is defined by a zone of sheared and brecciated host rock which grades into pipe fill breccia. The structure is surrounded

* Description based on Gornitz and Kerr (1970) and Chenoweth and Malan (1969).

Fig. 17. Major hydrothermal uranium districts in the western United States.

by a set of concentric circular fractures related to subsidence in the pipe. The diameter of the pipe varies from 40 m in the Hermit Shale to 150 m in the Supai Formation. The dimensions of the pipe, extent of alteration, and intensity of mineralization are controlled by wall rock lithology; in the Supai Formation all three features are developed to their maximum extent.

Pipe fill consists exclusively of downward displaced pieces of the sedimentary units traversed by the pipe. The breccia fragments range in size from 0.5 mm to several meters in diameter. The upper level of the pipe is filled mostly with loosely consolidated sandstone; the amount of brecciation increases downward. A zone of calcareous (carbonatized) sandstone surrounds the pipe. A similar carbonatized sandstone usually forms the matrix for the pipe breccia, although pyritic cement is important locally. Siltstone and shale breccia zones are concentrated near pipe margins. The center of the pipe contains massive non-brecciated sandstone blocks.

The Orphan uranium deposit is associated with the largest of several known mineralized and unmineralized sedimentary collapse structures in the Grand Canyon area and nearby Cameron district. This deposit combines many characteristics of both sandstone and hydrothermal uranium mineralization. Uranium ore occurs in fractures of the pipe border zone and also as irregularly shaped masses within the pipe. Uranium mineralization is most intense in the relatively porous sandstones of the breccia matrix. Siltstone and shale breccia fragments are generally unmineralized. The distribution of introduced minerals was locally controlled by faults and fractures.

Mineralogy and Paragenesis:

See Table 5 and Figure 18.

Notes: 1) Pitchblende is the only important primary uranium mineral; it occurs mostly as thin coatings on detrital quartz grains, but also in banded veins.

2) In high grade ore from the fractured border zone, pitchblende is invariably intergrown with red earthy hematite and sulfides; e.g., a core of pyrite or chalcopyrite is surrounded in sequence by concentric rings of pitchblende, a thin band of bleached rock, and a diffuse halo of hematite.

3) Pyrite is the most abundant sulfide.

4) Niccolite and rammelsbergite, present at the Orphan mine, have not been reported elsewhere on the Colorado Plateau.

5) Gangue minerals are calcite, dolomite, siderite and barite.

6) No vanadium minerals are present; an analysis of high grade uranium ore showed the presence of only 0.05% V_2O_5.

Table 5. Mineralogy of the Orphan mine (modified from Gornitz and Kerr, 1970).

ELEMENT	MINERAL	
	Hypogene	Supergene
Uranium	Uraninite	Torbernite Metatorbernite Zeunerite Metazeunerite Uranophane Bayleyite
Gold	Native gold	
Copper	Chalcocite Bornite Chalcopyrite Tennantite	Chalcocite Covellite Digenite Azurite Malachite Brochantite Native copper
Iron	Pyrite Siderite Marcasite Hematite	Limonite Ankerite Jarosite Hematite
Lead	Galena	Cerussite Anglesite
Manganese		Pyrolusite "Wad" Rhodochrosite
Molybdenum	Molybdenite	Wulfenite Ilsemannite
Nickel and cobalt	Skutterudite Siegenite Bravoite Rammelsbergite Niccolite	Erythrite Bieberite
Arsenic	Arsenopyrite	
Zinc	Sphalerite	Smithsonite
Magnesium	Dolomite	Hexahydrite Leonhardtite
Calcium	Calcite	Calcite Gypsum
Barium	Barite	

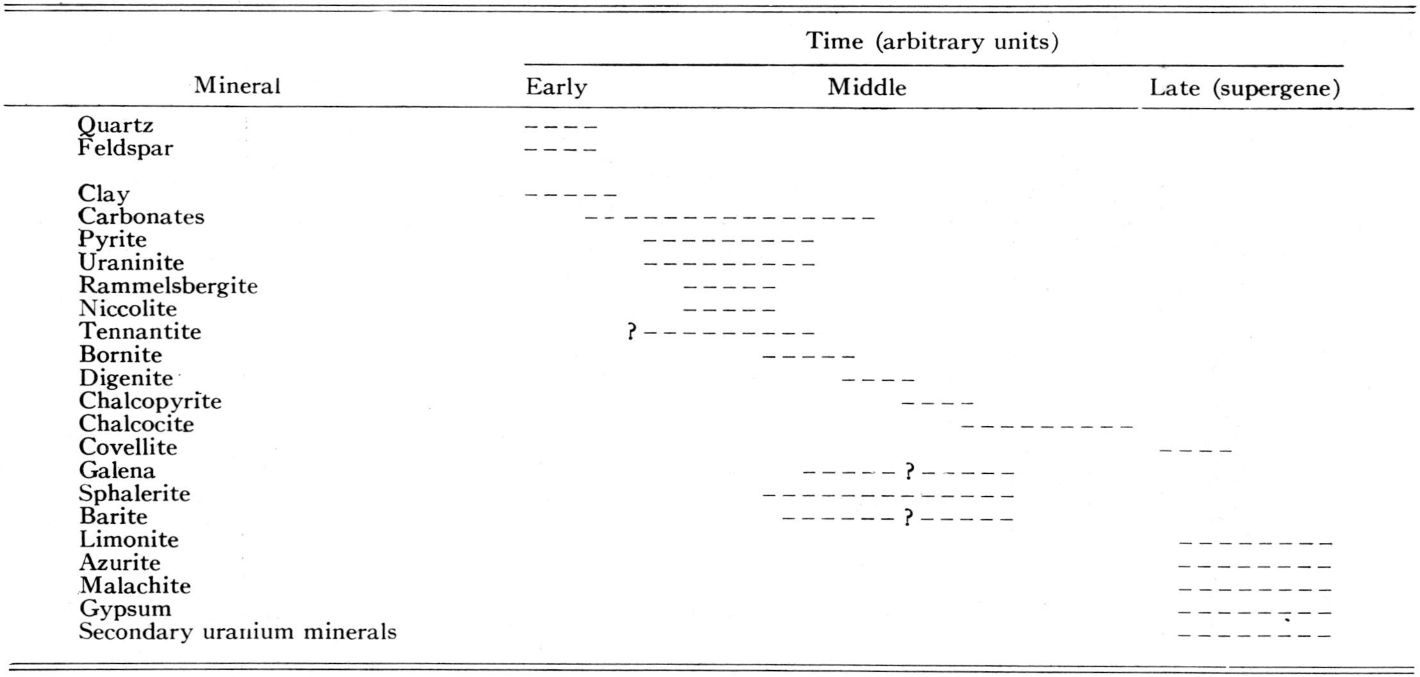

Mineral	Time (arbitrary units)		
	Early	Middle	Late (supergene)
Quartz	– – – –		
Feldspar	– – – –		
Clay	– – – – –		
Carbonates	– – –	– – – – – – – – – – – –	
Pyrite		– – – – – – – – –	
Uraninite		– – – – – – – – –	
Rammelsbergite		– – – – –	
Niccolite		– – – – –	
Tennantite		? – – – – – – – – –	
Bornite		– – – – –	
Digenite		– – – –	
Chalcopyrite		– – – –	
Chalcocite		– – – – – – – – –	
Covellite			– – – –
Galena		– – – – – ? – – – – –	
Sphalerite		– – – – – – – – – – – – – –	
Barite		– – – – – – ? – – – – –	
Limonite			– – – – – – – –
Azurite			– – – – – – – –
Malachite			– – – – – – – –
Gypsum			– – – – – – – –
Secondary uranium minerals			– – – – – – – –

Fig. 18. Paragenesis of the Orphan Mine (from Gornitz and Kerr, 1970).

Zoning:

1) Pyrite and pitchblende occur alone in the central part of the pipe and grade outward to a complex assemblage of pitchblende with sulfides and arsenides at the pipe margins.
2) The amount of pitchblende decreases downward, with large quantities present only above the 365′ level. The highest grade ore was present between the adit (0′) level and the 225′ level.
3) Copper mineralization occurs at lower levels than uranium (i.e. between the 400′ and 190′ levels).
4) Both pyrite and marcasite are found throughout the entire explored vertical extent (200m) of the pipe.

Wall Rock Alteration:

Wall rock alteration associated with the mineralization event is characterized primarily by bleaching of normally red sediments, carbonatization of originally non-calcareous sandstones, and the development of hematite halos around pitchblende concentrations. No difference was observed between the clay contents of mineralized and barren rocks. These effects are similar to alteration effects observed in classic Colorado Plateau uranium deposits.

1) Bleaching: Red bed fragments of the Supai Formation have been bleached within and near the contact of the pipe. In addition, tongues of bleached sediments extend several hundred feet outward from the pipe into permeable Supai beds. Unbleached Supai sediments contain 3-10% hematite.
2) Carbonatization is characterized by the development of coarse-grained calcite and dolomite cement and by the partial replacement of quartz grains by carbonate in pipe sandstones. Sandstones that were originally 90% quartz contain 30-50% carbonate after alteration.

Age:

The minimum age of uranium mineralization is 141 m.y. (Jurassic). The breccia pipe formed between the deposition of the Permian host rocks and the time of mineralization.

2) COLORADO

There are three important hydrothermal uranium districts in Colorado: the Front Range (Schwartzwalder mine), the Marshall Pass (Pitch mine), and the Cochetopa (Los Ochos mine) districts (Figure 17). Only the Schwartzwalder mine near Golden, Colorado has been a uranium producer of major importance.

a) Front Range Region

The hydrothermal deposits of the Ralston Creek and Golden Gate

Canyon areas (Figure 19) account for nearly all of the Front Range uranium production and reserves. Only minor amounts of sulfides and gangue are present in the pitchblende veins. Their wall rock is almost always Precambrian Idaho Springs Formation, a thick metamorphic sequence with an age of about 1.8 b.y. The uranium veins generally occur where major fault zones (breccia reefs) cut brittle, dark, mafic units of the Idaho Springs Formation. Figure 20 illustrates the localization of the deposits of the Ralston Creek and the Golden Gate Canyon areas by major fault zones. Figure 23 shows the effect of lithologic control in the area.

(1) Schwartzwalder Mine*

Location:

Front Range, Ralston Creek, 13 km northwest of Golden, Colorado, 105° 20′ W, 39° 40′ N (see Figure 20).

Geology:

See Figure 21 and Plate I of Sheridan et al. (1967).

The country rock of most of the Schwartzwalder mine area is Precambrian, consisting of complexly folded and faulted metasedimentary (and metavolcanic?) rocks. Metamorphic rock types include biotite-quartz-plagioclase gneiss, mica schist, microcline-quartz-plagioclase-biotite gneiss, amphibolite, hornblende gneiss, calc-silicate gneiss, quartzites, quartzitic gneisses, and impure marbles. In the mine itself, the important rock units are quartz-biotite schist and gneiss (locally garnetiferous), lime silicate-hornblende gneiss, quartzite, and pegmatite. Most ore occurs within the garnetiferous biotite gneiss and adjacent quartz-biotite schist. Some ore also occurs in clean quartzites. Calc-silicate rocks in the mine are usually barren; ore veins tend to pinch out when they enter this unit. The biotitic gneiss and schist contain sparsely distributed graphite flakes.

Less than 1 km northeast of the Schwartzwalder mine, Precambrian rocks are overlain unconformably by a sequence of Upper Paleozoic and Mesozoic units, beginning with the red sandstones of the Pennsylvanian Fountain Formation. Laramide dikes and sills are found within a few kilometers of the mine. Northwest trending Laramide faulting has produced systems of large breccia reefs and fracture zones; the mineralization of the Schwartzwalder mine is developed in subsidiary structures related to one of these zones, the Rogers Breccia Reef.

Ore at the Schwartzwalder consists of pitchblende-mineralized

*Description based on the authors' observations and the published accounts of Sheridan et al. (1967), Downs and Bird (1965), Heyse (1971) and Young and Lahr (1975).

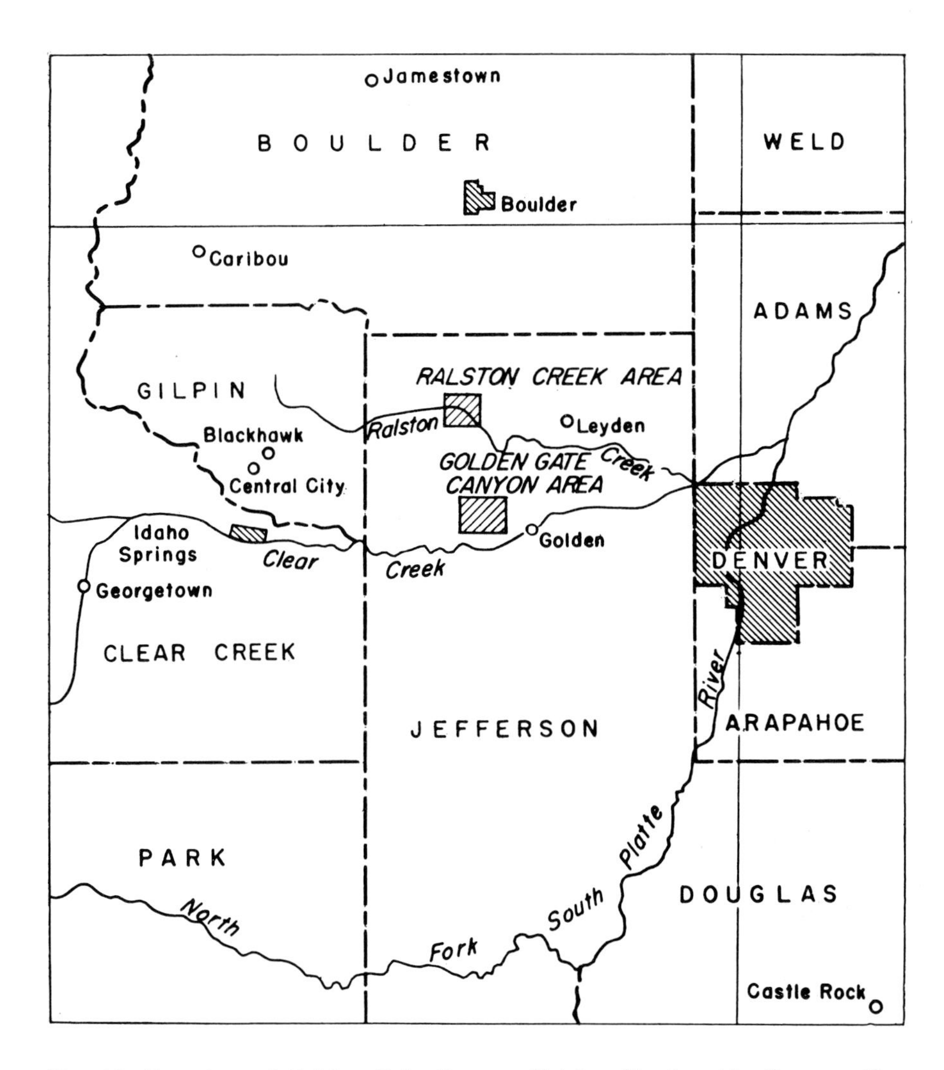

Fig. 19. Location of Golden Gate Canyon, Ralston Creek, and other uraniferous districts, Front Range, Colorado (from Adams and Stugard, 1956).

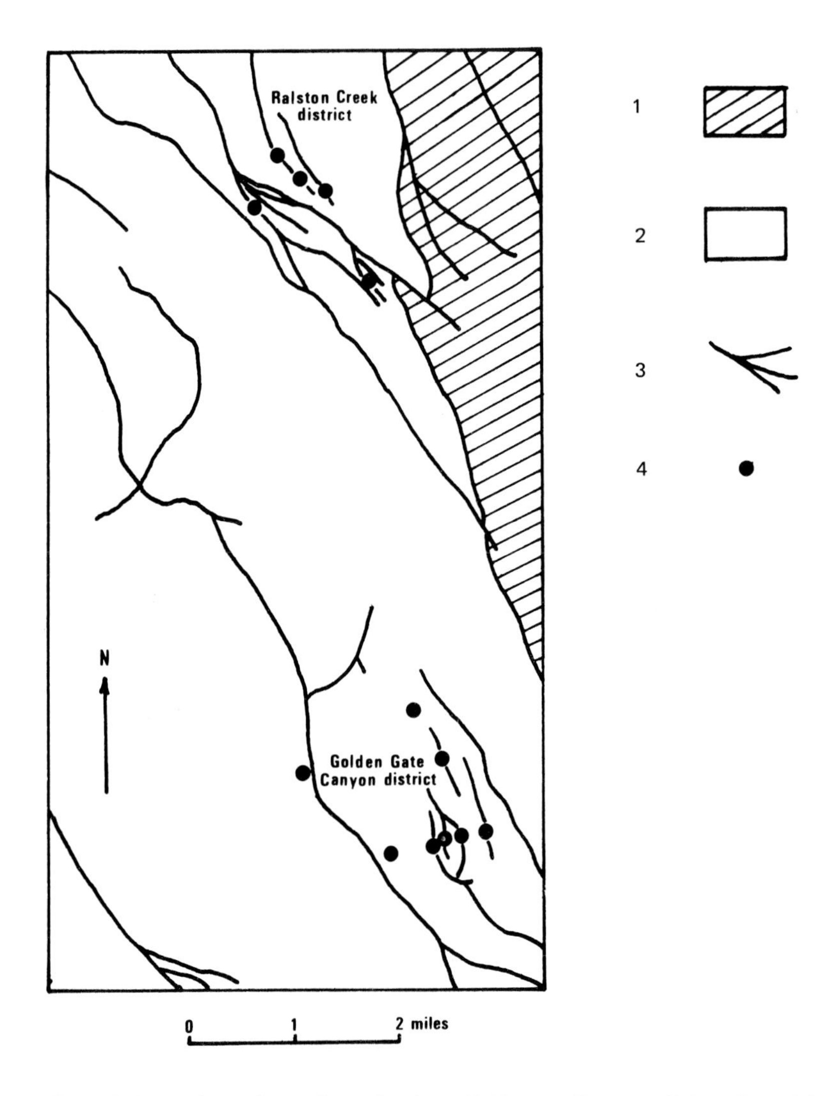

Fig. 20. Loxation of uranium districts, Jefferson County, Colorado, with respect to major faults (redrawn from Sheridan et al., 1958). 1 = Mesozoic and Paleozoic sedimentary rocks; 2 = Precambrian crystalline rocks; 3 = fault; 4 = Uranium deposit.

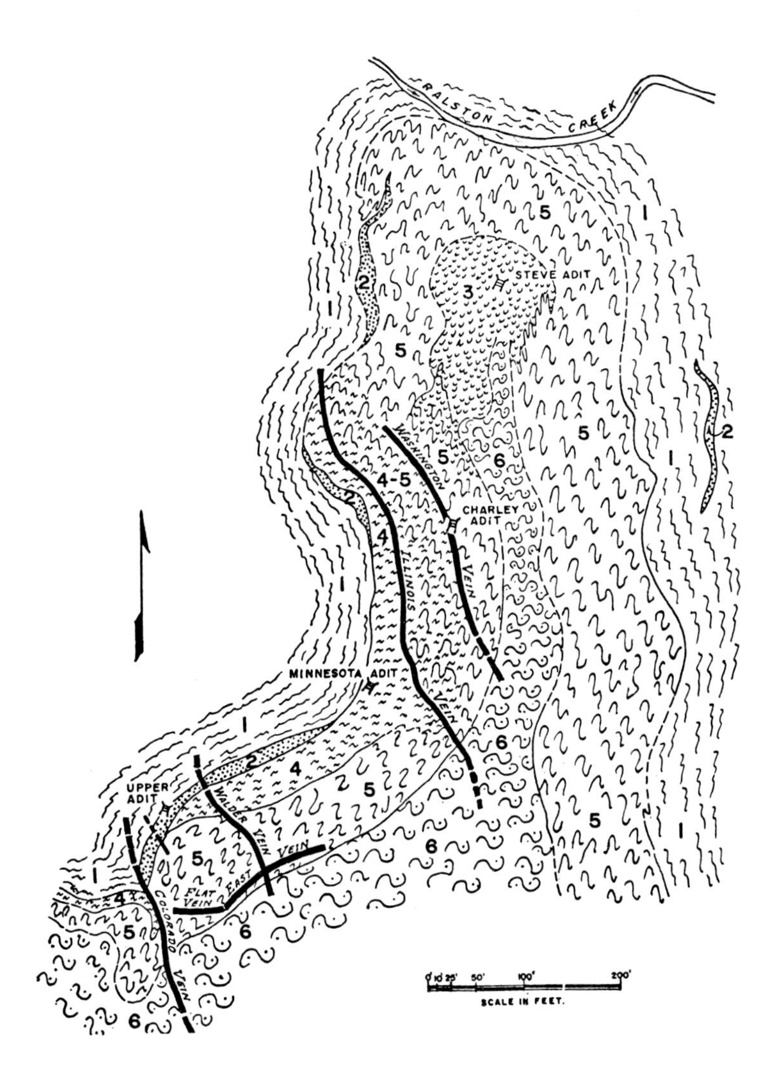

Fig. 21. Surface geology of the Schwartzwalder mine showing the distribution of units of the Idaho Springs Formation: unit 1 = lime silicate — hornblende schist; unit 2 = metaquartzite; unit 3 = chlorite schist; unit 4 = phyllitic schist; unit 5 = garnetiferous biotite gneiss; unit 6 = quartz-rich muscovite schist. The location of the portals of the adits and the outcrop of the principal veins are also indicated (from Downs and Bird, 1965).

fractures and breccia zones. Pitchblende is frequently found coating red, altered wall rock fragments. Uranium ore occurs throughout the entire explored vertical distance of more than 700 m. Wall rocks of the mine frequently contain disseminated pyrite and pyrrhotite. These sulfides are found most abundantly in the ore-bearing garnetiferous transition zone that marks the contact between the hornblende gneiss and mica schist units. The disseminated pyrite and pyrrhotite are probably not related to the uranium mineralization, but to pre-ore metamorphic processes. Sulfides in the wall rocks, however, could have served as reducing agents for later pitchblende deposition.

Mineralogy and Paragenesis:

See Table 6 and Figure 22.

Vein mineralogy is characterized by the occurrence of pitchblende (with rare coffinite), pyrite/marcasite, and minor amounts of jordisite (amorphous molybdenum sulfide), and base metal sulfides in a gangue of ankerite, adularia, calcite, and quartz.

Table 6. Vein Mineralogy of the Schwartzwalder Mine

Major	Minor	Trace
adularia	chalcopyrite	barite
ankerite	coffinite	bornite
calcite	galena	emplectite
hematite	jordisite	niccolite
marcasite	quartz	sylvanite
pitchblende	siderite	
pyrite	sphalerite	
	tennantite	

Notes:
1) The dominant gangue mineral is ankerite.
2) Red, hematitic radioactive breccia and pitchblende vein aureoles are frequently observed.
3) The presence of adularia as a vein mineral in the Schwartzwalder ore has been verified by X-ray and optical techniques. It occurs as tiny euhedral crystals and aggregates of anhedral grains. Adularia commonly coats vein walls and breccia fragments. It is the only mineral that is clearly intergrown with pitchblende.
4) No vanadium minerals are known to be present, but analyses reported by Sheridan et al. (1967) indicate a content of as much as 0.077% V_2O_5 in the ore.

Wall Rock Alteration:

Hematization is the most striking wall rock alteration effect present at the Schwartzwalder mine. Other alteration effects present in the mine

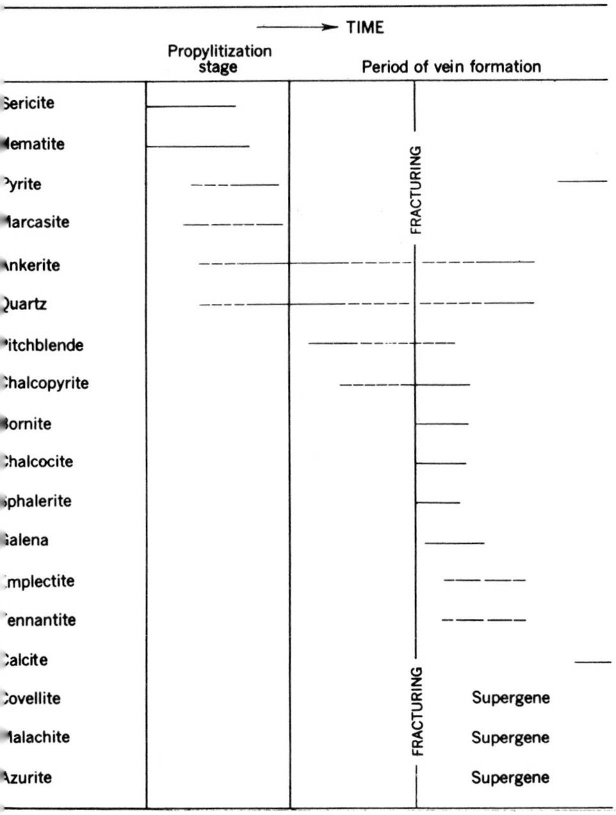

TIME

Propylitization stage

Period of vein formation

FRACTURING

Sericite
Hematite
Pyrite
Marcasite
Ankerite
Quartz
Pitchblende
Chalcopyrite
Bornite
Chalcocite
Sphalerite
Galena
Emplectite
Tennantite
Calcite
Covellite — Supergene
Malachite — Supergene
Azurite — Supergene

FRACTURING

Fig. 22. Paragenesis of the Schartzwalder mine (from Walker et al., 1963).

include carbonatization, feldspathization and chloritization. Alteration appears to be confined mainly to veins, fault breccias and to coarsely brecciated wall rocks that contain pitchblende. Breccia fragments of wall rock are generally altered to a reddish, fine-grained aggregate of ankerite, quartz and adularia. Locally, breccias are silicified. Zones of bleached wall rock surround some fault zones, breccias, and veins; where present, these bleached zones frequently occur between unaltered wall rocks and the more intensely altered red aureoles which are found immediately adjacent to veins and breccia zones.

Age:

The mineralized structures are of Laramide age. Vein pitchblendes yield Late Cretaceous-Early Tertiary isotopic ages: 73 ± 5 m.y. (Sheridan et al., 1967); 65 m.y. (Young and Lahr, 1975).

Ore Control:

Uranium mineralization is concentrated in favorable structural sites (e.g. faults, fractures, and fault zone breccias) and in favorable lithologic units (e.g. rocks rich in hornblende, biotite, and garnet).

(2) Union Pacific Mine*

Location:

Front Range, Golden Gate Canyon, Colorado, 105° 20′ W, 39° 40′ N (Figures 19 and 23).

Geology:

See Figure 23 and Plate I of Sheridan et al. (1967).

Country rocks consist of interlayered hornblende gneiss, biotite schist, and quartz-biotite gneiss.

Mineralogy:

Pitchblende, chalcopyrite, tetrahedrite-tennantite, bornite, chalcocite, covellite, pyrite, ankerite, potash feldspar, emplectite, sphalerite, galena, chlorite, leucoxene, hematite, calcite, malachite, and azurite.

Paragenesis:

See Figure 24.

Wall Rock Alteration:

1. Propylitization producing chlorite and sericite.
2. Potassic alteration.

Mineralized zone is intensely altered.

* Description based on Walker et al. (1963) and Adams and Stugard (1956).

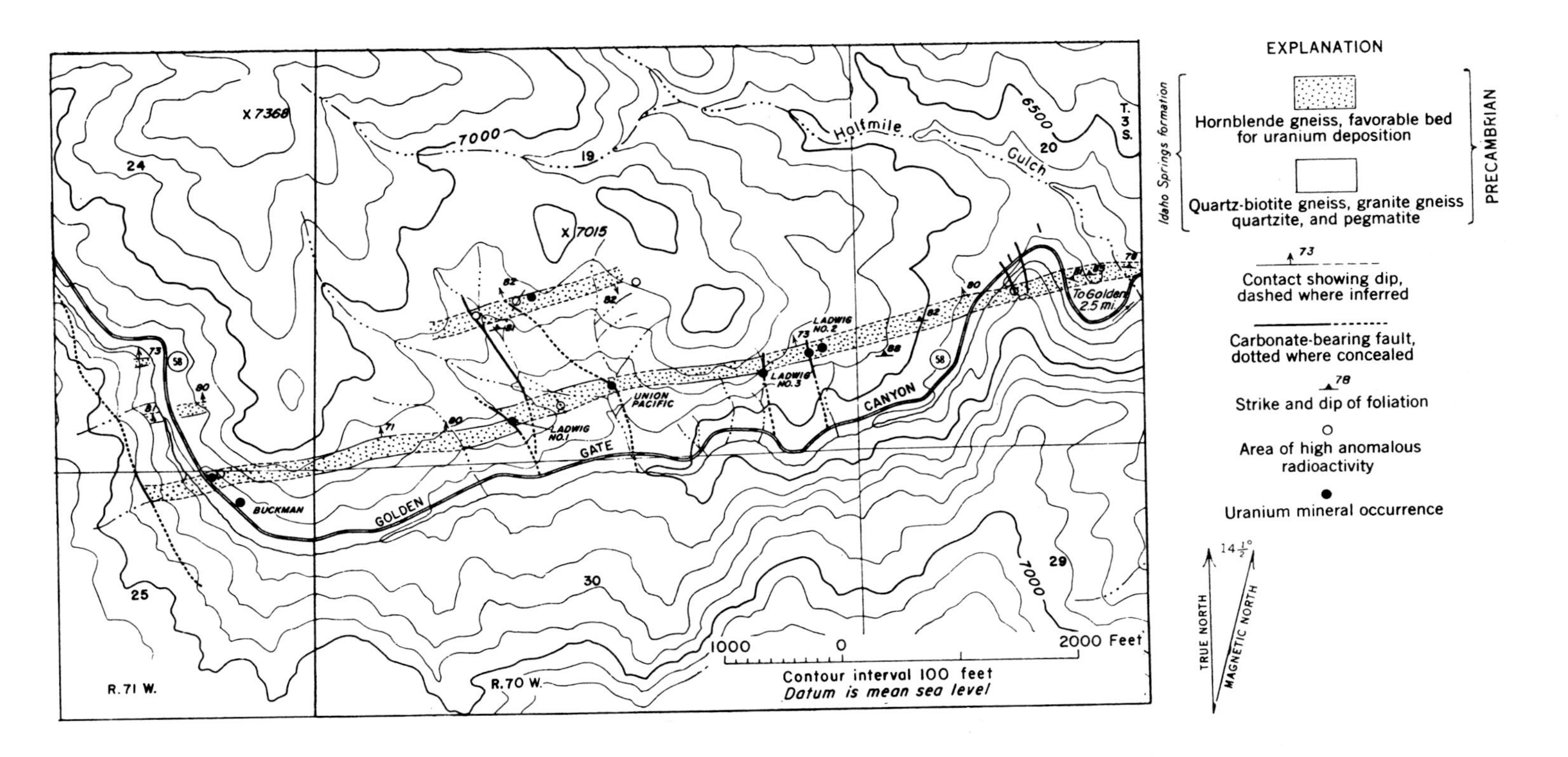

Fig. 23. Geologic sketch map showing the distribution of uranium occurrences in Golden Gate Canyon area, Colorado (from Adams and Stugard, 1956).

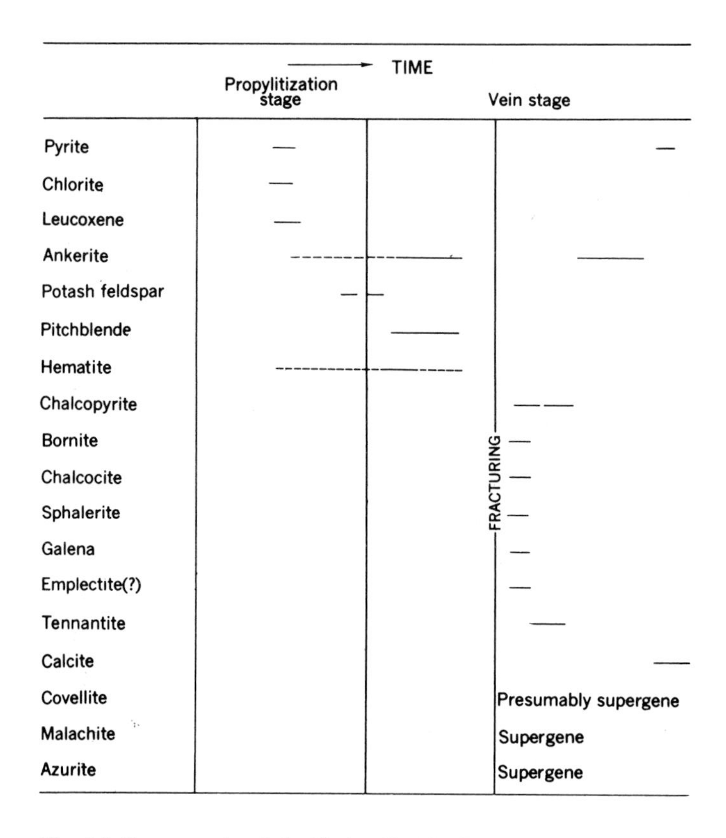

Fig. 24. Paragenesis of the Union Pacific Prospect (from Walker et al., 1963).

Age:

The faults containing uranium mineralization are of Laramide age.

Ore Controls:

The mineralization occurs at the intersection of north trending carbonate-bearing fault zones and a thin, N 75° E trending belt of hornblende gneiss and biotite schist. Uranium ore is confined to the narrow band of hornblende gneiss.

(3) Caribou Mine*

Location:

Front Range, Boulder County, Colorado, 105° 35′ W, 39° 58′ N (Figure 19).

Geology:

Ore veins occur chiefly in the monzonite of the Caribou stock. This stock is a composite Laramide body which has intruded Precambrian biotite gneiss.

Mineralogy:

Gold, silver, pitchblende, galena, pyrite, hematite, sphalerite, chalcopyrite, tetrahedrite, gersdorffite, argentite, proustite, cobalt, copper and manganese oxides, fluorite, cerargyrite, pyrargyrite, azurite, malachite, cerussite and barite.

Quartz, dolomite, and calcite are common gangue minerals.

Paragenesis:

Stage A:
1. Quartz + calcite + siderite.
2. Pyrite.
3. Chalcopyrite.
4. Sphalerite
5. Galena.

Stage B:
1. Gersdorffite + chalcedony.
2. Pitchblende + chalcedony + minor pyrite.
3. Sphalerite + chalcopyrite + minor pyrite and pitchblende.
4. Pyrite.
5. Argentite + chalcopyrite.
6. Proustite followed by minor pitchblende.
7. Native silver.

* Description based on Moore et al. (1957), Wright (1954) and Smith (1938).

Wall Rock Alteration:

Silicification, carbonatization, sericitization, argillization, and chloritization effects have been reported. Wall rock alteration aureoles surrounding veins can be several meters thick.

Age:

The host monzonite and associated intrusives are of Late Cretaceous or Tertiary age; the uranium mineralization is thought to be Tertiary.

(4) Central City District*

Location:

Front Range, Gilpin and Clear Creek Counties, Colorado, 39° 48′ N, 105° 30′ W (Figure 19).

Geology:

The country rocks consist of Precambrian granite gneiss, schist, and calc-silicate rock. These rocks are cut by Tertiary granite, quartz monzonite, granodiorite and bostonite. The bostonite dikes are 2 to 5 times more radioactive than any of the other intrusive rock types in the district; locally they contain as much as 100 ppm uranium.

Mineralogy:

Gold, silver, pitchblende, galena, pyrite, hematite, sphalerite, chalcopyrite, tetrahedrite-tennantite, torbernite, metatorbernite, autunite, kasolite, dumontite, enargite, bismuth, molybdenite, bornite, gold tellurides, marcasite, chalcocite, and covellite.

The gangue is predominantly quartz, but locally carbonates (calcite, ankerite, siderite and rhodocrosite), fluorite and barite are important.

Paragenesis:

See Figures 25 and 26.

The sporadic uranium mineralization of the Central City district is apparently unrelated to the economically more important sulfide ores. Veins of the pyrite and base metal sulfide stages are closely related and show a regional zonation. Uranium mineralization predates the sulfide stages and does not conform to the sulfide zoning. Where pitchblende and sulfide stages occur in the same veins, the sulfides appear to have been deposited in re-opened uranium veins.

* Description based on Walker et al. (1963), Sims and Barton (1962), and Everhart (1956).

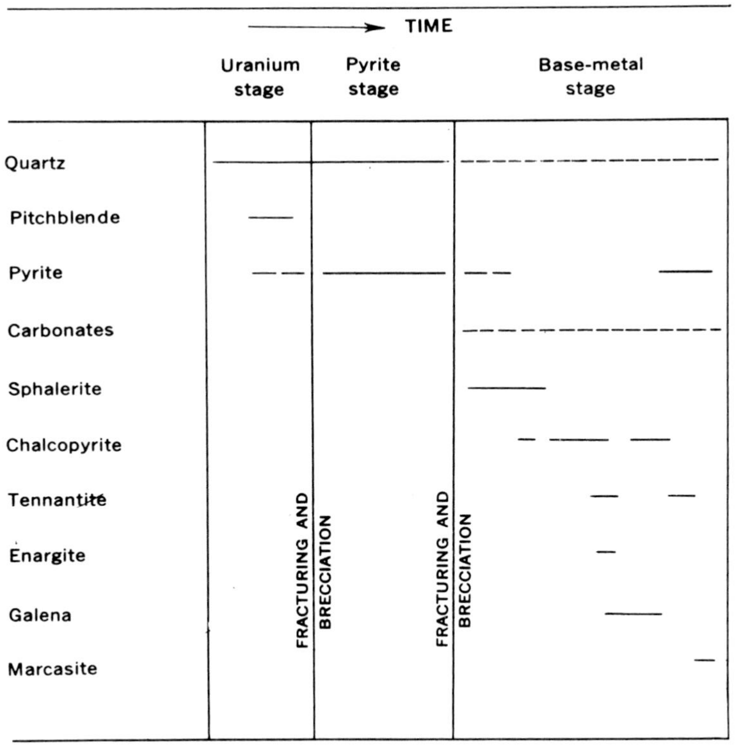

TIME

Uranium stage | Pyrite stage | Base-metal stage

Quartz

Pitchblende

Pyrite

Carbonates

Sphalerite

Chalcopyrite

Tennantite

Enargite

Galena

Marcasite

FRACTURING AND BRECCIATION

FRACTURING AND BRECCIATION

Fig. 25. Generalized paragenesis of the vein minerals of the Central City district (from Walker et al., 1963).

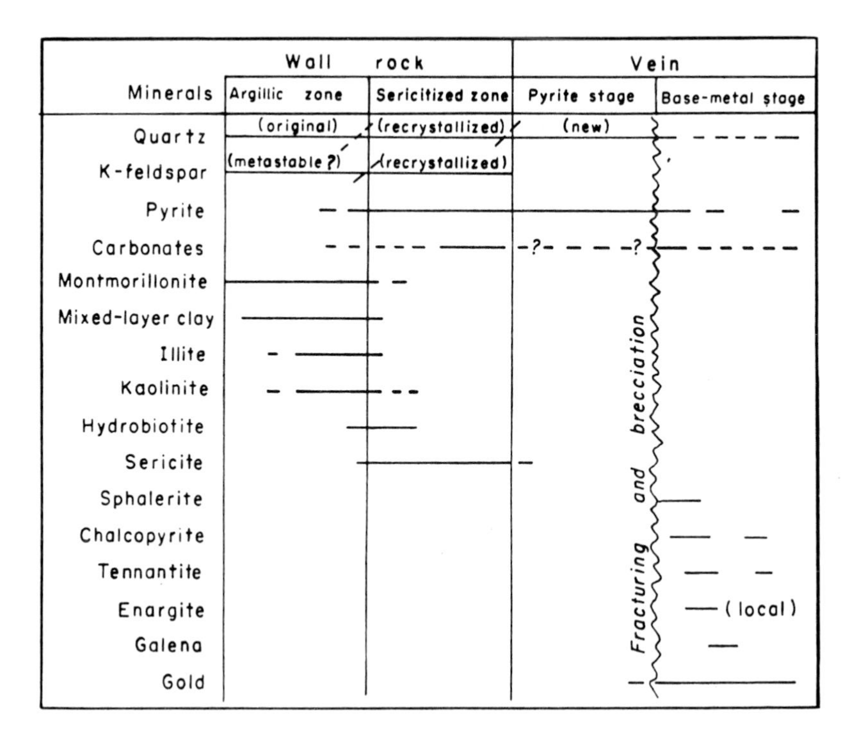

Fig. 26. Generalized paragenesis of vein and wall rock alteration minerals, Central City district (from Sims and Barton, 1962; reprinted from the Buddington Volume, Geological Society of America).

Zoning:

A core area (2 miles in diameter) of quartz-pyrite veins is surrounded by intermediate and peripheral zones in which copper ores pass outward to lead-zinc ores. The distribution of pitchblende mineralization cross-cuts the pattern of sulfide zoning.

Wall Rock Alteration:

See Figure 26.

Vein walls are silicified, sericitized, and pyritized in a zone several cm thick adjacent to the vein. This hard, sericitized zone grades outward into a strongly argillized zone, and finally into a weakly argillized zone. The alteration zones of the pyrite-quartz veins and composite pyrite-base metal veins are thicker than those of veins containing only base metal sulfides.

Age:

Walker et al. (1963) report isotopic ages of 55-70 m.y. for pitchblende from the Central City district.

Ore Controls:

Ore veins follow zones of minor faulting and brecciation; changes in the trend or dip of a vein are loci of mineralization. Thick mineralized veins occur in competent quartz monzonite; veins are thin and barren in gneiss. Uranium mineralization is spatially related to the occurrence of highly radioactive bostonite dikes.

b) Marshall Pass District

Pitch Mine*

Location:

Marshall Pass district, Sawatch Range, Saguache County, Colorado, 106° 14′ W, 38° 23′ N (Figure 17).

Geology:

See Figure 27.

The Pitch mine is the most important hydrothermal uranium occurrence in the Marshall Pass region. In this area faulting and erosion have reduced the Paleozoic sequence to isolated remnants which unconformably overlie the quartz-biotite-potassium feldspar schists, gneissic granites, and pegmatites of the Precambrian basement. One such remnant, consisting of Pennsylvanian and older sedimentary rocks, is bounded on the east by the Chester reverse fault (Laramide?). The uranium deposits of the Marshall Pass district are found along this fault. To the south, the

* Description based on Chenoweth and Malan (1969), except as otherwise noted.

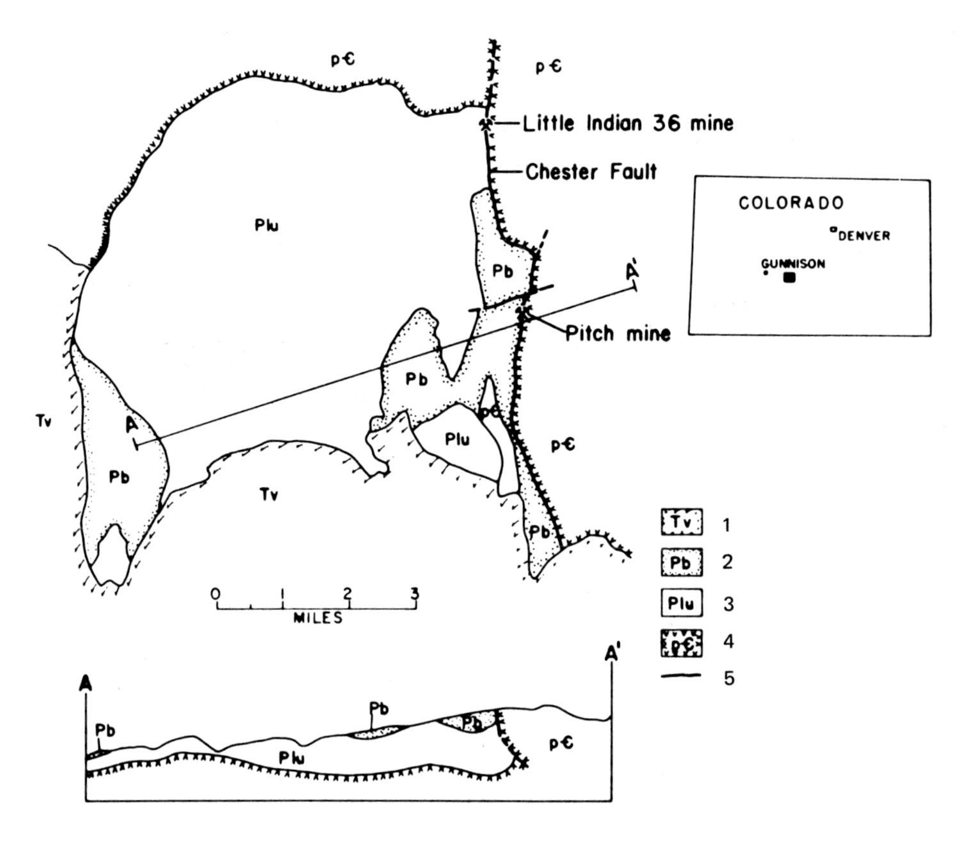

Fig. 27. Generalized geology of the Marshall Pass area, Colorado (from Chenoweth and Malan, 1969). 1 = Tertiary volcanics; 2 = Paleozoic Belden Formation; 3 = Lower Paleozoic undivided; 4 = Precambrian; 5 = fault.

Paleozoic outlier is overlain by Miocene volcanic flows; to the west, north and east Precambrian basement is exposed.

At the Pitch mine, uranium mineralization occurs in fault-controlled veins and breccia zones in limestone of the Pennsylvanian Belden Formation. The host limestone is interbedded with lignite, shale, and fine-grained sandstone. Ore is found in the footwall of the Chester fault; Precambrian crystalline rocks compose the hanging wall. Ore shoots occur where wedges of Precambrian rocks have been displaced from the hanging wall of the Chester fault to the footwall zone along second order tensional faults. Pitchblende is the chief ore mineral of the mine, but traces of copper and iron sulfides also occur. Gross (1965), in his description of a similar uranium deposit from the Marshall Pass region, states that the uranium vein and immediately adjacent wall rocks are hematized.

c) Cochetopa District

Los Ochos Mine*

Location:

Cochetopa district, Saguache County, Colorado, 106° 50′ W, 38° 20′ N (Figure 17).

Geology:

See Figure 28.

The country rocks of the Los Ochos mine area consist of Precambrian quartz-biotite schist, hornblende gneiss, and minor ultrabasic rocks, which are intruded by granitic bodies and overlain by Mesozoic sedimentary rocks and Tertiary volcanics. Miocene intrusives are also present in the area. The Mesozoic sediments include Jurassic sandstones and mudstones (Morrison Formation) and Cretaceous sandstones (Dakota Formation) and shales (Mancos Formation). The Los Ochos uranium deposit occurs in brecciated and silicified Morrison sediments along a high-angle normal fault.

Mineralogy:

Pitchblende is the primary uranium mineral; it is associated with abundant clay and minor marcasite. Uranophane, autunite, torbernite, ilsemannite, and chalcocite are also present. The gangue consists of clays, quartz, chalcedony, and barite. Sulfide mineralization was accompanied by alunite.

* Description based on Chenoweth and Malan (1969), Malan and Ranspot (1959), and Derzay (1956).

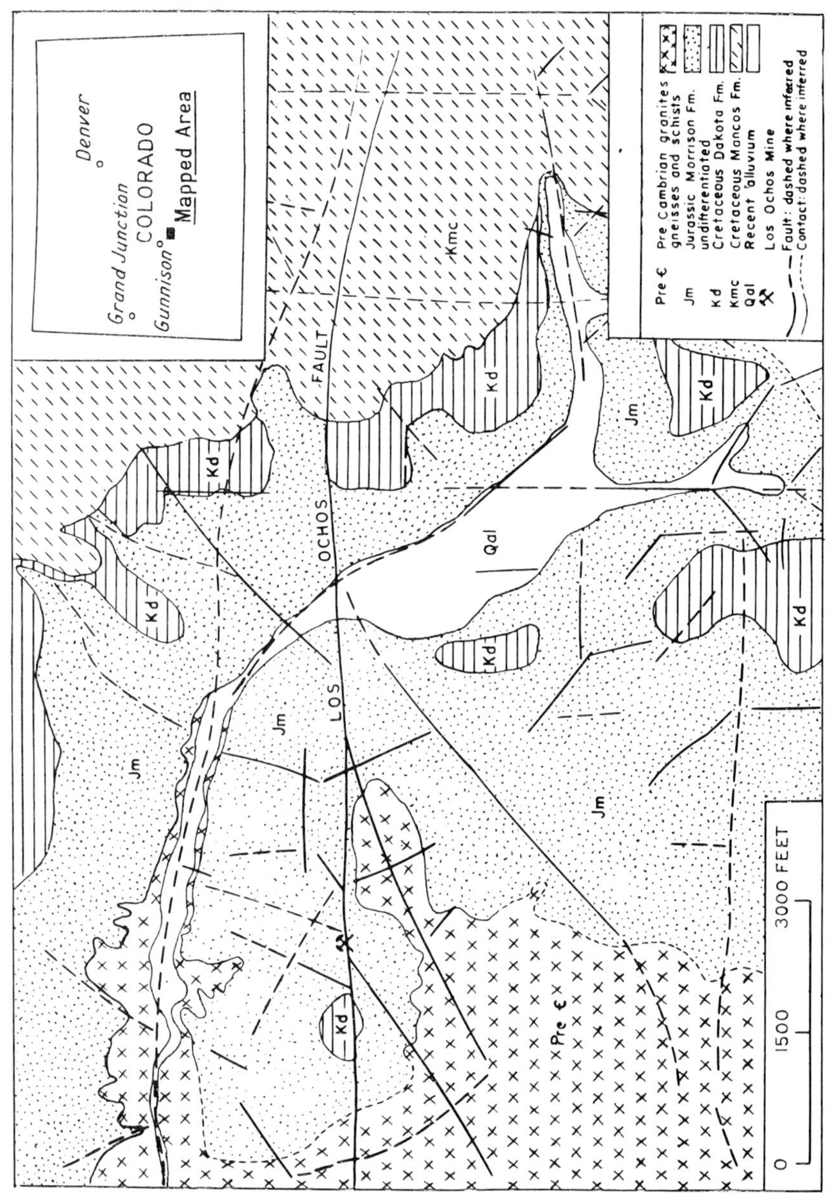

Fig. 28. Surface geology of the Los Ochos mine area, Colorado (from Kerr, 1956; United Nations).

Paragenesis:

1) Tertiary faulting.
2) Silicification.
3) Fracturing, brecciation, and hydrothermal alteration.
4) Marcasite I + clays.
5) Refracturing.
6) Marcasite II.
7) Pitchblende.
8) Kaolinite in late fracture fillings.

Wall Rock Alteration:

Hydrothermal alteration is intense in the vicinity of ore bodies and has affected a zone several hundred meters thick along the Los Ochos fault. Silicification and kaolinization are predominant, but chloritization and sericitization are also important in the altered basement rocks. Locally, the red and green variegated mudstones of the Morrison Formation are bleached.

Age:

Uranium mineralization post-dates Tertiary faulting.

Ore Controls:

The uranium mineralization occupies a thick shear zone at the contact between Precambrian granite and Jurassic sediments. Where this contact cuts the Los Ochos fault there is a large pipe-like zone of mineralization. Silicification and brecciation of Morrison and Dakota sedimentary rocks in and adjacent to fault zones controlled uranium deposition.

3) IDAHO

Sunshine Mine*

Location:

Coeur d'Alene district, Idaho, 116° 45′ W,47° 40′ N.

Geology:

Country rocks: St. Regis quartzite and argillite of the Precambrian Belt Series, intruded by Laramide monzonite stocks. Minor hydrothermal uranium mineralization pre-dates the economically important silver and base metal veins of the deposit.

*Description based on Fryklund (1960), except as otherwise noted.

Mineralogy:

See Table 7.

The gangue is predominantly siderite and quartz. Pyrite and tetrahedrite are the most abundant sulfides. Uraninite veins contain more pyrite and arsenopyrite than copper-silver ores; when uranium increases, silver decreases. Uraninite is often found along vein margins and within the adjacent wall rocks.

Table 7. Mineralogy of the Sunshine Mine, Idaho

allophane	galena
arsenopyrite	gersdorffite
azurite	hematite
barite	jasper
boulangerite	malachite
bournonite	Mn-oxides
cerargyrite	proustite
cerussite	pyrite
chalcocite	pyrrhotite
chalcopyrite	quartz
chrysocolla	siderite
covellite	sphalerite
cuprite	stibnite
erythrite	tetrahedrite
Fe-oxides	uraninite

Paragenesis:

Generalized sequence:

1) Precambrian: a) Disseminated arsenopyrite + pyrite.
b) Bleaching of wall rocks.
c) Uraniferous veins (pyrite + uraninite + quartz + hematite).

2) Late Cretaceous and Tertiary: Main period of silver-base metal mineralization (contains no uranium).

Wall Rock Alteration

Bleaching and sericitization (zones up to 700 m thick) are common. Silicification is related to pitchblende deposition. Carbonatization is related to silver mineralization. Hematization of the St. Regis quartzite is uniquely associated with uraniferous veins.

Age:

Uraninite: 710 ± 10 m.y. (Kerr and Kulp, 1952).

Ore Controls:

Brecciation and fracture cleavage provided structural sites for pitchblende deposition. Arsenopyrite-bearing ores are favorable for uranium mineralization (Anderson, 1940). Uranium occurs only in the Precambrian St. Regis Formation and not in the Laramide intrusives.

4) NEVADA

Reese River (Austin) District*

Location:

Near Austin, Lander County, Nevada, 117° 7′ W, 39° 28′ N (Figure 17).

Geology:

Uranium mineralization is found at the southern edge of the Reese River district, spatially unrelated to the important silver ores. The district centers about an intrusive quartz monzonite body, which is probably Jurassic in age. Uranium mineralization occurs in the intensely fractured contact zone between the quartz monzonite and the Precambrian or Cambrian quartzite and argillite host rocks. Thin mineralized fracture zones cut the quartz monzonite, but the metasedimentary rocks are the more favorable ore host. Uranium mineralization occurs as disseminations and fault and fracture fillings in small roof pendants of metasediments near the intrusive contact. In general appearance, the uranium veins resemble dikes and are composed chiefly of fine-grained sericite and varying amounts of quartz. Autunite and metatorbernite were the chief uranium minerals mined in the district. Pitchblende has not been reported, but probably occurs at depth.

5) OREGON

Lakeview Deposit**

Location:

Lake County, south-central Oregon, 120° 25′ W, 42° 27′ N (Figure 17).

Geology:

Hydrothermal uranium deposits occur in the contact zone between a late Tertiary rhyolite intrusive and Miocene pyroclastics and lacustrine sediments. Uraninite and secondary uranium minerals occur as veinlets, irregular masses, and disseminations in stockworks, which have been modified by faulting and groundwater redistribution.

*Description based on Thurlow (1956) and Chenoweth and Malan (1969).

**Description from Chenoweth and Malan (1969).

6) UTAH

Marysvale District*

Location:

Central Utah, 112° 21′ W, 38° 30′ N (Figures 17 and 29).

Geology:

See Figures 29 and 30 and the maps accompanying Kerr et al. (1957).

The Marysvale district is located near the eastern margin of the Basin and Range Province in a terrane dominated by Tertiary volcanics. The volcanics are intruded by small bodies of quartz monzonite and related granitic rocks. Uranium mineralization generally occurs as vein and replacement deposits in intrusive quartz monzonite or granite. Replacement bodies of uranium mineralization are sometimes found in rhyolite adjacent to quartz monzonite containing uranium veins. Uranium ores are known over a vertical range of more than 600 m, but most of the mined ore was oxidized and came from within about 150 m of the present erosion surface.

Mineralogy and Paragenesis:

See Table 8 and Figure 31.

Notes: 1) Pyrite and radioactive purple to black fluorite are closely associated with pitchblende in veins and breccia.

2) Some opal is radioactive and fluorescent.

Table 8. Mineralogy of the Marysvale District, Utah

adularia	johannite
allophane	jordisite
alum	kaolinite
alunite	leucoxene
autunite	limonite
barite	magnetite
biotite	montmorillonite
beta-uranotile	nontronite
calcite	opal
celadonite	phosphuranylite
chalcedony	penninite
clinochlore	pitchblende
dickite	pyrite
fluellite	quartz
fluorite	rutile
gearksutite	schroeckingerite
goethite	sericite
gypsum	torbernite
halite	tyuyamunite

* Except as otherwise noted, this description is based on the authors' observations and on the published account of Kerr (1968).

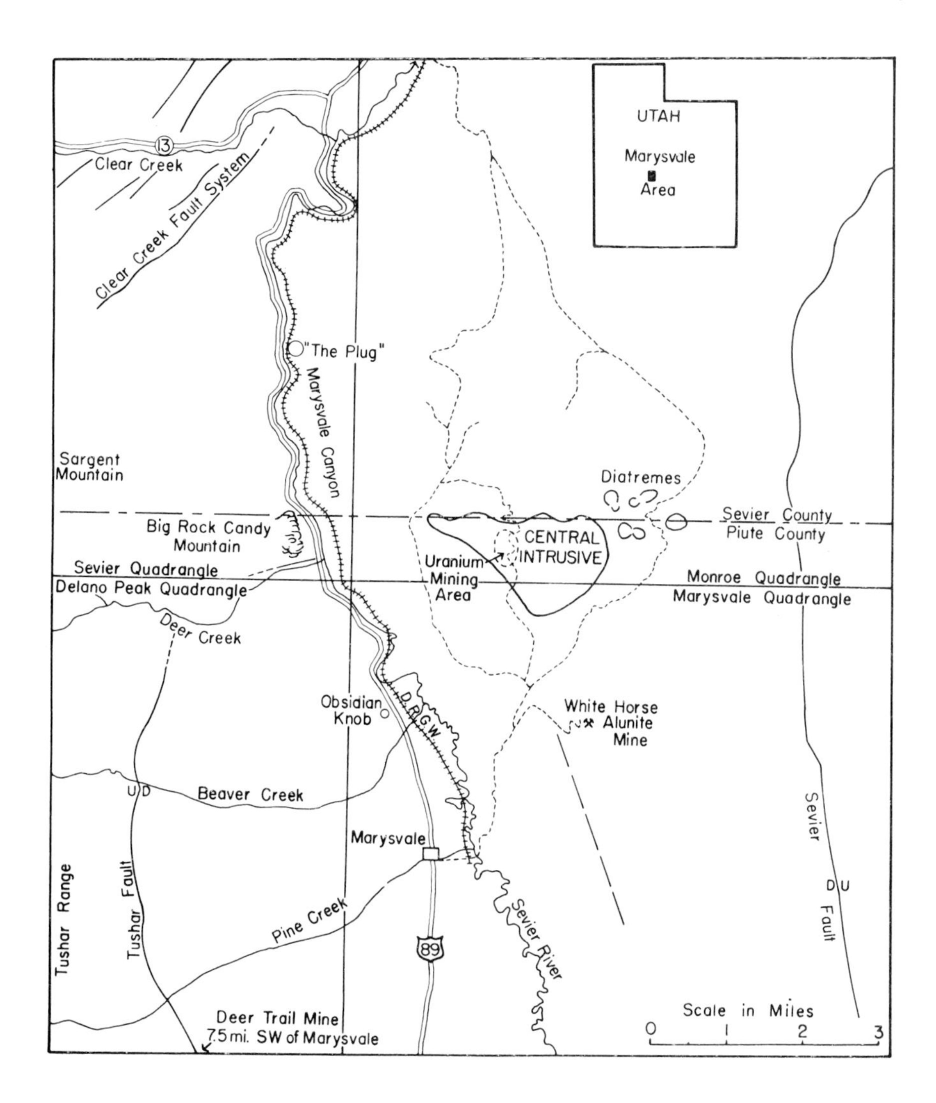

Fig. 29. Locality map for the Marysvale district, Utah (from Kerr, 1963).

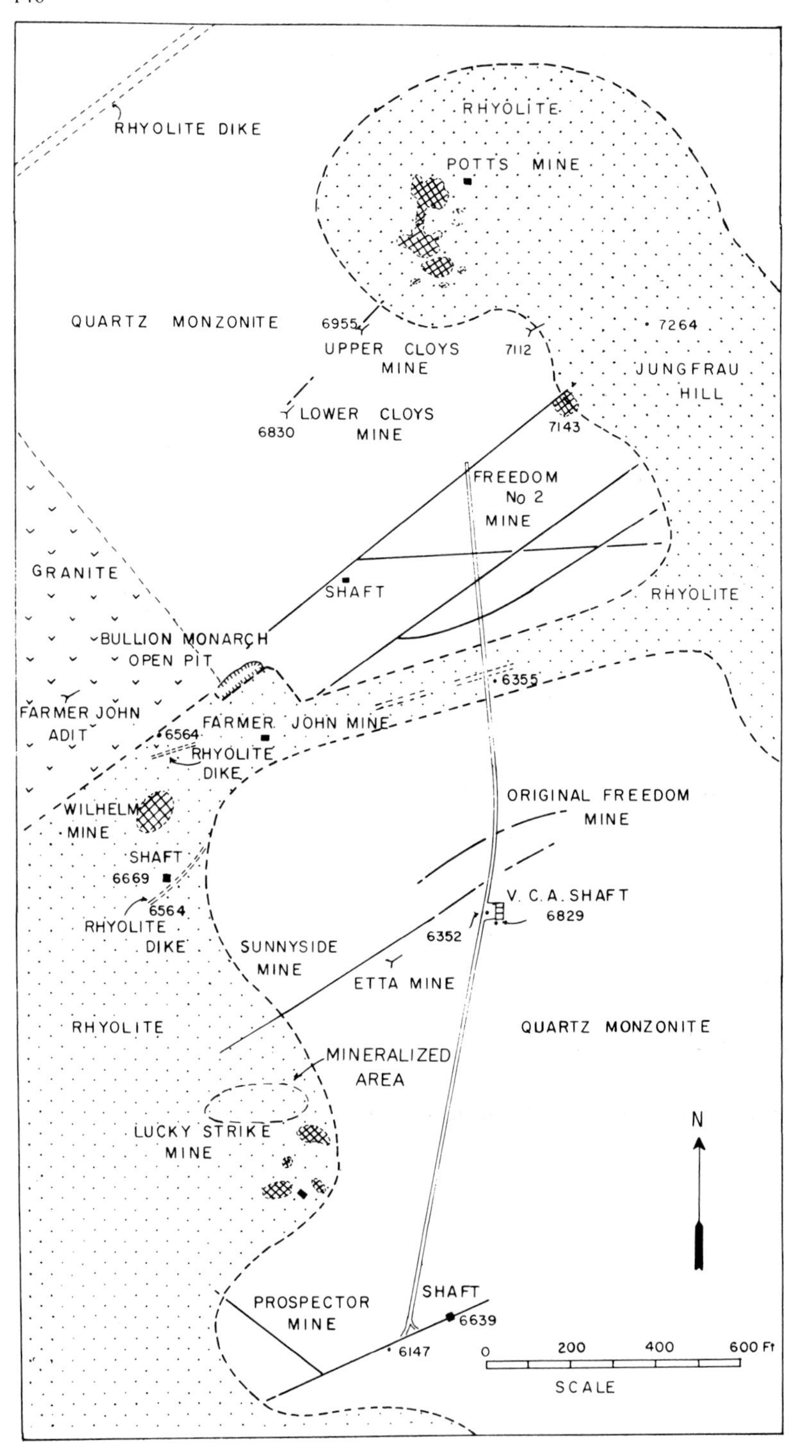

Fig. 30. Geology of the Marysvale uranium district. Shallow workings in rhyolite are cross-hatched. Surface projections of the principal veins are shown as black lines (from Kerr, 1968).

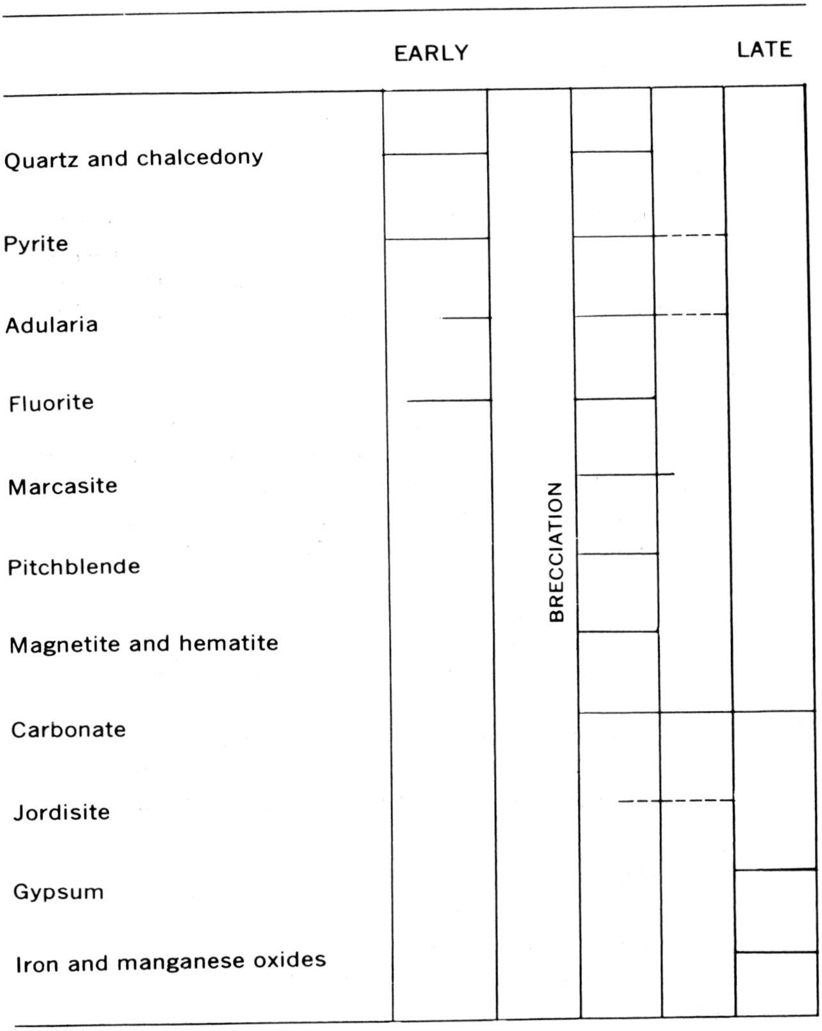

EARLY LATE

Quartz and chalcedony

Pyrite

Adularia

Fluorite

Marcasite

Pitchblende

Magnetite and hematite

Carbonate

Jordisite

Gypsum

Iron and manganese oxides

BRECCIATION

Fig. 31. Paragenesis of the Marysvale district (from Walker et al., 1963).

hematite	umohoite
halloysite	uranophane
illite	uranopilite
ilsemannite	zippeite
jarosite	

Zoning:

Uranium occurs in the center of the district, surrounded by alunite deposits. It is possible that the alunitic alteration may form a cap above the less intense argillic alteration zones that contain the uranium mineralization. Alunite does not occur more than 100 m below the present surface.

Wall Rock Alteration:

According to Kerr et al. (1957) there are two major periods of hydrothermal alteration in the Marysvale district. The first is the pervasive, pre-ore alunitic alteration; the second is the moderate to intense argillic alteration associated with the deposition of pitchblende.

Alteration zones associated with Marysvale uranium veins:

Unaltered rock.

1) Some alteration of ferromagnesian minerals.
2) Gray-green chloritic band; calcite veinlets.
3) Gray-white argillic band; sericitization; some fluorite veinlets; chlorite replaced by clay minerals.
4) White argillic band; fluorite present.

Vein: pitchblende + black fluorite + pyrite + fine-grained quartz + sericite + calcite + molybdenum minerals.

Ages:

Pitchblende: 10 to 13 m.y.
Rhyolite: 17.5 m.y. (average)
Quartz monzonite: 23 to 26 m.y.
Bullion Canyon Volcanics: 28 to 31 m.y.

Ore Controls:

1. Northeast trending faults and fractures.
2. Brecciated zones of previously silicified rock.
3. Quartz monzonite is the favored host rock.

7) WASHINGTON

Midnite Mine*

Location:

35 miles northwest of Spokane, Washington, 18° 5′ W, 47° 55′ N (Figure 17).

Geology:

See Figure 32.

The Midnite mine area is underlain by Precambrian Togo Formation, pelitic and calcareous metasedimentary rocks, which have been intruded by a Cretaceous porphyritic quartz monzonite pluton that contains an average of 12 ppm uranium. At the mine, steeply dipping metasediments occur as a roof pendant above and adjacent to the quartz monzonite intrusion. Essentially all primary uranium ore is found in the metasedimentary rocks, close to the intrusive contact. The contact zone is highly fractured.

Pelitic host rocks are graphitic phyllites and schists which consist of quartz + muscovite + andalusite ± biotite ± pyrite. Calcareous hornfels and marble host rocks are composed of calcite ± diopside ± phlogopite ± idocrase ± garnet ±scapolite ± plagioclase ± pyrite. In addition, amphibolite sills and Tertiary dacite dikes occur in the mine area and locally contain ore.

Pitchblende is found as disseminations along foliation, replacements, and fracture fillings. Uranium ore bodies cross-cut sedimentary lithologic boundaries without noticeable effect; the ore bodies are pitchblende stockworks or irregular mineralized masses which are generally elongated parallel to the intrusive contact and often occupy topographic lows in this contact zone (Figure 33).

Mineralogy:

See Table 9.

Below the oxidized zone, pitchblende and some coffinite occur with pyrite and marcasite, the most abundant sulfides. In addition, small amounts of pyrrhotite and molybdenite, and traces of arsenopyrite, chalcopyrite and sphalerite occur at the Midnite mine. The inconspicuous gangue includes quartz, carbonate, and iron oxides. Smoky quartz is commonly observed in the intrusive rock.

*Description based on the authors' observations and on the published accounts of Nash and Lehrman (1975) and Barrington and Kerr (1961).

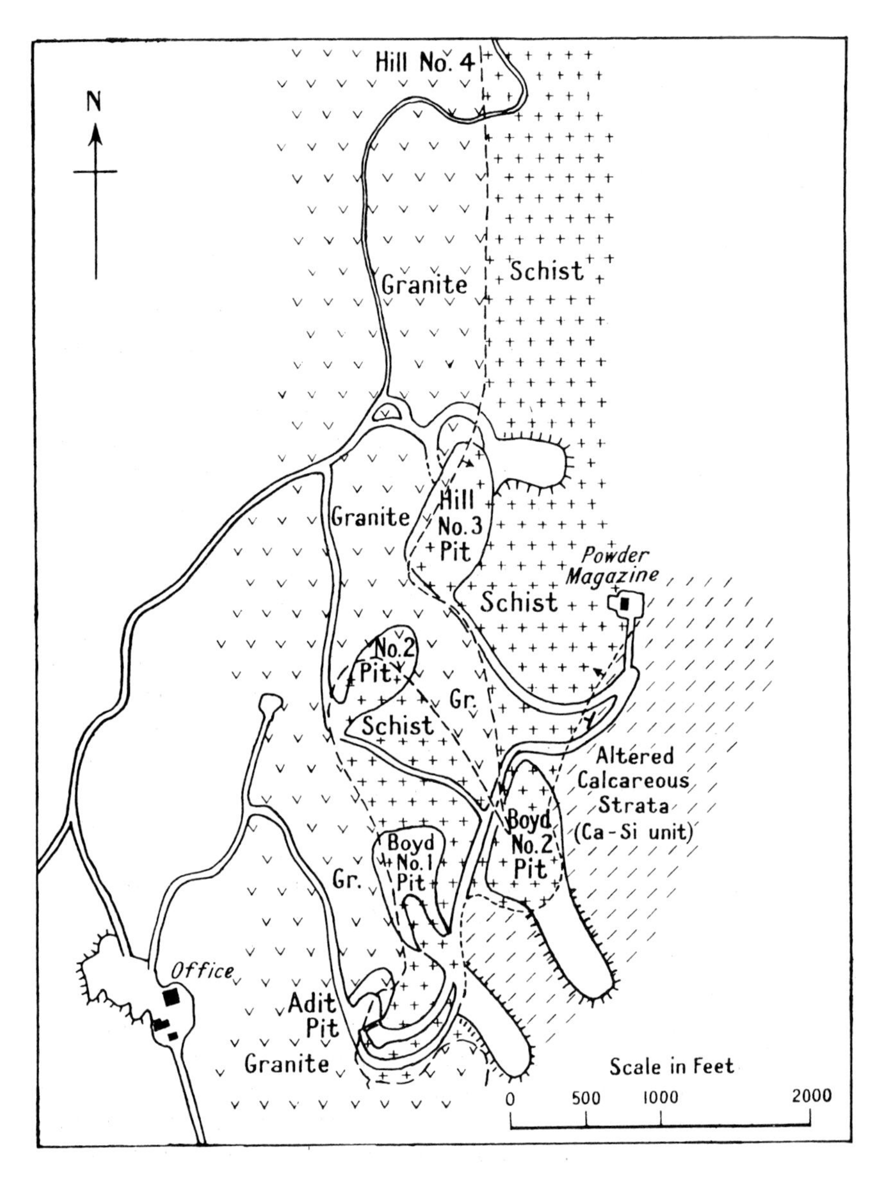

Fig. 32. Geology of the Midnite mine (from Barrington and Kerr, 1961).

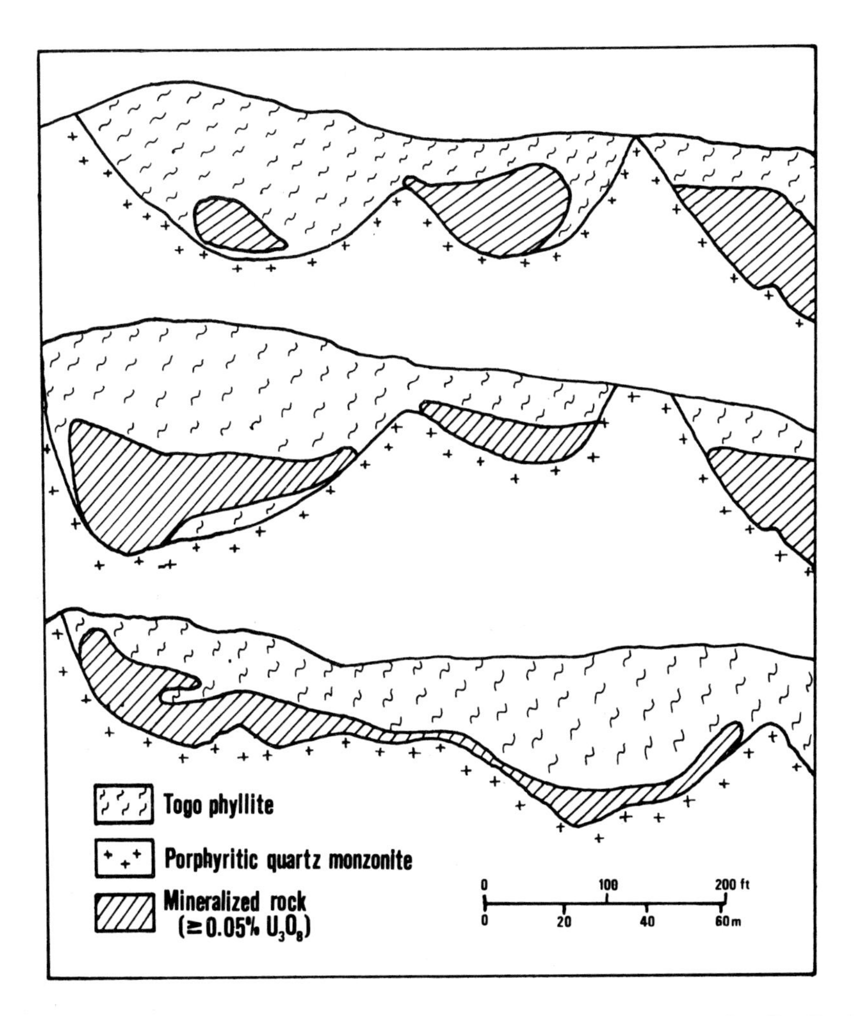

Fig. 33. Geologic cross-sections of the Midnite mine showing the distribution of uranium mineralization (redrawn from Nash and Lehrman, 1975).

Table 9. Mineralogy of the Midnite Mine, Washington

adularia	meta-autunite
apatite	molybdenite
arsenopyrite	pitchblende
autunite	pyrite
chalcopyrite	phosphuranylite
clay minerals	rutile
coffinite	sklodowskite
delvauxite	sphalerite
gummite	torbernite
hematite	tourmaline
liebigite	uraninite
magnetite	uranophane
marcasite	zippeite

Paragenesis:

Generalized sequence:

1. Emplacement of granitic rocks.
2. Molybdenite, followed by arsenopyrite + pyrite + quartz.
3. Argillic alteration producing illite and kaolinite (relation to Stage 2 is not known).
4. Marcasite + pitchblende + minor chalcopyrite and sphalerite.
5. Montmorillonite produced in wall rock during and after uranium mineralization.

Wall Rock Alteration:

Host rocks at the Midnite mine are reported to be bleached, argillized, sericitized, kaolinized and chloritized, but ore-associated alteration is not prominent.

Age:

The isotopic age of pitchblende mineralization is 102 to 108 m.y. Lead-alpha ages of zircon from the porphyritic quartz monzonite and the granodiorite indicate that the Loon Lake pluton is Cretaceous in age.

Ore controls:

1) The fractured intrusive contact between Precambrian metasediments and the Cretaceous porphyritic quartz monzonite.
2) Ore favors irregularities in the granite-metasediment contact such as noses, embayments, and changes of strike and dip.

Selected United States References

Adams, J.W. and Stugard, F., 1956, Wall rock control of certain pitchblende deposits in Golden Gate Canyon, Jefferson County, Colorado: U.S. Geol. Survey Bull. 1030-G, 187-209.

Anderson, R.J., 1940, Microscopic features of ore from the Sunshine mine: Econ. Geology, *35,* 659-667.

Barrington, J. and Kerr, P.F., 1961, Uranium mineralization at the Midnite mine, Spokane, Washington: Econ. Geology, *56,* 241-258.

Becraft, G.E. and Weis, P.L., 1963, Geology and mineral deposits of the Turtle Lake quadrangle, Washington: U.S. Geol. Survey Bull. 1131, 73 pp.

Butler, A.P., Jr., Finch, W.I. and Twenhofel, W.S., 1962, Epigenetic uranium deposits in the United States, exclusive of Alaska and Hawaii: U.S. Geol. Survey Mineral Inv. Resource Map MR-21.

Chenoweth, W.L. and Malan, R.C., 1969, Significant geologic types of uranium deposits - United States and Canada: Remarks at U.S. Atomic Energy Comm. Uranium Workshop, Grand Junction, Colorado, Paper No. 5.

Cooper, M., 1953, Bibliography and index of literature on uranium and thorium and radioactive occurrences in the U.S.; Part 1—Arizona, Nevada, and New Mexico: Geol. Soc. America Bull., *64*, 197-234.

Cooper, M., 1953, Bibliography and index of literature on uranium and thorium and radioactive occurrences in the U.S.; Part 2—California, Idaho, Montana, Oregon, Washington, and Wyoming: Geol. Soc. America Bull., *64*, 1103-1172.

Cooper, M., 1954, Bibliography and index of literature on uranium and thorium and radioactive occurrences in the U.S.; Part 3—Colorado and Utah: Geol. Soc. America Bull., *65*, 467-590.

Cooper, M., 1955, Bibliography and index of literature on uranium and thorium and radioactive occurrences in the U.S.; Part 4—Arkansas, Iowa, Kansas, Louisiana, Minnesota, Missouri, Nebraska, North Dakota, Oklahoma, South Dakota, and Texas: Geol. Soc. America Bull., *66*, 257-326.

Curtis, D., 1958, Selected annotated bibliography of the uranium geology of igneous and metamorphic rocks in the U.S.: U.S. Geol. Survey Bull. 1059-E, 205-262.

Dean, B.G., 1960, Selected annotated bibliography of the geology of uranium-bearing veins in the U.S.: U.S. Geol. Survey Bull. 1059-G, 327-440.

Derzay, R.C., 1955, Geology of the Los Ochos uranium deposit, Saquache County, Colorado: U.S. Geol. Survey Prof. Paper 300, 137-141.

Downs, G.R. and Bird A.G., 1965, The Schwartzwalder uranium mine, Jefferson County, Colorado: Mtn. Geologist, *2*, no. 4, 183-191.

Drake, A.A., Jr. 1957, Geology of the Wood and East Calhoun mines, Central City district, Gilpin County, Colorado: U.S. Geol. Survey Bull. 1032-C, 129-170.

Everhart, D.L., 1956, Uranium-bearing vein deposits in the United States: Internat. Conf. Peaceful Uses of Atomic Energy, *6*, 257-264, United Nations.

Fryklund, V.C., Jr., 1960, Ore deposits of the Coeur d'Alene district, Shoshone County, Idaho: U.S. Geol. Survey Prof. Paper 445, 103 pp.

Gornitz, V. and Kerr, P.F., 1970, Uranium mineralization and alteration, Orphan mine, Grand Canyon, Arizona: Econ. Geology, *65*, 751-768.

Gross, E.B., 1965, A unique occurrence of uranium minerals, Marshall Pass, Saguache County, Colorado: Am. Mineralogist, *50*, 909-923.

Heyse, J.V., 1971, Mineralogy and paragenesis of the Schwartzwalder mine uranium ore, Jefferson County, Colorado: U.S. Atomic Energy Comm. Rept. GJO-912-1, 91 pp.

Kerr, P.F., 1956, The natural occurrence of uranium and thorium: Internat. Conf. Peaceful Uses of Atomic Energy, *6*, 5-59, United Nations.

Kerr, P.F., 1963, Geologic features of the Marysvale uranium area, Utah: Intermtn. Assoc. Petroleum Geologists, 12th Ann. Field Conf. Guidebook, 125-135.

Kerr, P.F., 1968, The Marysvale, Utah, uranium deposits: in Ridge, J.D., ed., *Ore Deposits of the United States, 1933-1967, 2*, 1020-1042, Am. Inst. Min. Met. and Pet. Engineers, New York.

Kerr, P.F., and Kulp, J.L., 1952, Pre-Cambrian uraninite, Sunshine mine, Idaho: Science, *115*, No. 2978, 86-88.

Kerr, P.F., and Robinson, R.F., 1953, Uranium mineralization in the Sunshine mine, Idaho: Am. Inst. Min. and Met. Engineers Trans., *196*, 495-511.

Kerr, P.F., Brophy, G.P., Dahl, H.M., Green, J. and Woolard, L.E., 1957, Marysville, Utah, uranium area: Geol. Soc. America Spec. Paper 64, 212 pp.

Malan, R.C. and Ranspot, H.W., 1959, Geology of the uranium deposits in the Cochetopa mining district, Saguache and Gunnison Counties, Colorado: Econ. Geology *54*, 1-19.

Merewether, E.A., 1960, Geologic map of the igneous and metamorphic rocks showing location of uranium deposits: U.S. Geol. Survey Misc., Geol. Inv. Map I–309, 1: 500,000.

Molloy, M.W. and Kerr, P.F., 1962, Tushar uranium area, Marysvale, Utah: Geol. Soc. America Bull., *73*, 211-236.

Moore, F.B., Cavender, W.S. and Kaiser, E.P., 1957, Geology and uranium deposits of the Caribou area, Boulder County, Colorado: U.S. Geol. Survey Bull. 1030-N, 517-550.

Nash, J.T. and Lehrman, N., 1975, Geology of the Midnite uranium mine, Stevens County, Washington—a preliminary report (abst.): Geol. Soc. America Abstracts with Programs, *7*, 634-635 (full text in U.S. Geol. Survey Open-file Report 75-402, 36 pp.).

Ross, C.P., 1953, The geology and ore deposits of the Reese River district, Lander County, Nevada: U.S. Geol. Survey Bull. 997, 132 pp.

Schafer, M., 1955, Preliminary report on the Lakeview uranium occurrences, Lake County, Oregon: Ore Bin, *17*, no. 12, 93-94, Oregon Dept. Geol. and Min. Ind., Portland.

Schnabel, R.W., 1955, The uranium deposits of the United States: U.S. Geol. Survey Mineral Inv. Resource Map MR-2, 1:500,000.

Sharp, B.J. and Hetland, D.L., 1954, Preliminary report on uranium occurrences in the Austin area, Lander County, Nevada: U.S. Atomic Energy Comm. Rept. RME-2010, 13 pp.

Sheridan, D.M., Maxwell, C.H. and Albee, A.L., 1967, Geology and uranium deposits of the Ralston Buttes district, Jefferson County, Colorado: U.S. Geol. Survey Prof. Paper 520, 121 pp.

Sheridan, D.M.: Maxwell, C.H., Albee, A.L. and Van Horn, R., 1958, Preliminary map of bedrock geology of the Ralston Buttes quadrangle, Jefferson County, Colorado: U.S. Geol. Survey Mineral Inv. Field Studies Map MF-179

Sims, P.K., 1956, Paragenesis and structure of pitchblende-bearing veins, Central City district, Gilpin County, Colorado: Econ. Geology, *51*, 739-756.

Sims, P.K. and Barton, P.M., 1962, Hypogene zoning and ore genesis, Central City district, Colorado: in Engel, A.E.J. and Leonard, B.F., eds., *Petrologic Studies:* A volume in honor of A.F. Buddington, 373-395, Geol. Soc. America.

Smith, W., 1938, Geology of the Caribou stock in the Front Range, Colorado: Am. Jour. Sci., *236* 161-196.

Soister, P.E. and Conklin, D.R., 1959, Bibliography of U.S. Geological Survey reports on uranium and thorium—1942 through May 1958: U.S. Geol. Survey Bull. 1107-A, 1-167.

Thurlow, E.E., 1955, Uranium deposits at the contact of metamorphosed sedimentary rocks and granitic intrusive rocks in western United States: U.S. Geol. Survey Prof. Paper 300, 85-89.

U.S. Atomic Energy Commission, 1959, Guidebook to uranium deposits of the western United States.

Walker, G.W., Osterwald, F.W. and Adams, J.W., 1963, Geology of uranium-bearing veins in the conterminous United States: U.S. Geol. Survey Prof. Paper 455, 146 pp.

Walthier, T.N., 1955, Uranium occurrences of the eastern United States: Mining Eng., 7, 545-547.

Weissenborn, A.E. and Moen, W.S., 1974, Uranium in Washington: in Livingston, V.E., Jr. et al., *Energy Resources of Washington*, Washington Div. Mines and Geology Inf. Circ. No. 50, 83-97.

Wright H.D., 1954, Mineralogy of a uraninite deposit at Caribou, Colorado: Econ. Geology, *49*, 129-173.

Young, E.J. and Lahr, M., 1975, The Schwartzwalder uranium mine, Jefferson County, Colorado (abst.): Geol. Soc. America Abstracts with Programs, 7, 653.

II) AUSTRALIA

The locations of major Australian hydrothermal uranium deposits are shown in Figure 34.

A) Darwin Region

The Darwin region of the Northern Territory (Figure 34) contains numerous pitchblende deposits. These deposits are concentrated in the Rum Jungle, South Alligator River, and Alligator Rivers areas. Dodson et al. (1974) state that, with few minor exceptions, all known uranium deposits of the Darwin region occur in Lower Proterozoic sedimentary rocks of the Pine Creek geosyncline. There appears to be a strong stratigraphic control on the distribution of these deposits; the uranium deposits of all three districts are found in lithologically similar and stratigraphically correlative units. These units are the Golden Dyke and Koolpin Formations, and the metamorphosed equivalent of the Koolpin Formation (Figure 35). For the detailed geology of the Darwin region, see the maps accompanying the report of Walpole et al. (1968).

1) Rum Jungle District*

Location:

Darwin region, Northern Territory, 13° S, 131° E (Figure 34).

Geology:

See Figure 36.

Country rocks consist of a Lower Proterozoic, low grade metasedimentary sequence composed of greywackes, arkoses, sandstones, conglomerates, dolomites, phyllites, and schists, which contains inliers of Archean crystalline basement (Rum Jungle Complex). The Archean basement contains 2-30 ppm uranium depending on the rock type. Granites average 10 ppm uranium and 46 ppm thorium. The uranium deposits occur in carbonaceous shales and chloritic slates of the Lower Proterozoic Golden Dyke Formation near outcrops of granitic basement. Ore lodes occupy shears, faults and tight folds.

Mineralogy:

See Table 10.

Pitchblende is the only important uranium ore mineral, and quartz is the most abundant gangue mineral. Chalcopyrite and pyrite are common. Magnetite and fluorite occur as accessory minerals in the metasediments.

*Description based on Dodson et al. (1974), Crohn (1968) and Roberts (1960), except as otherwise noted.

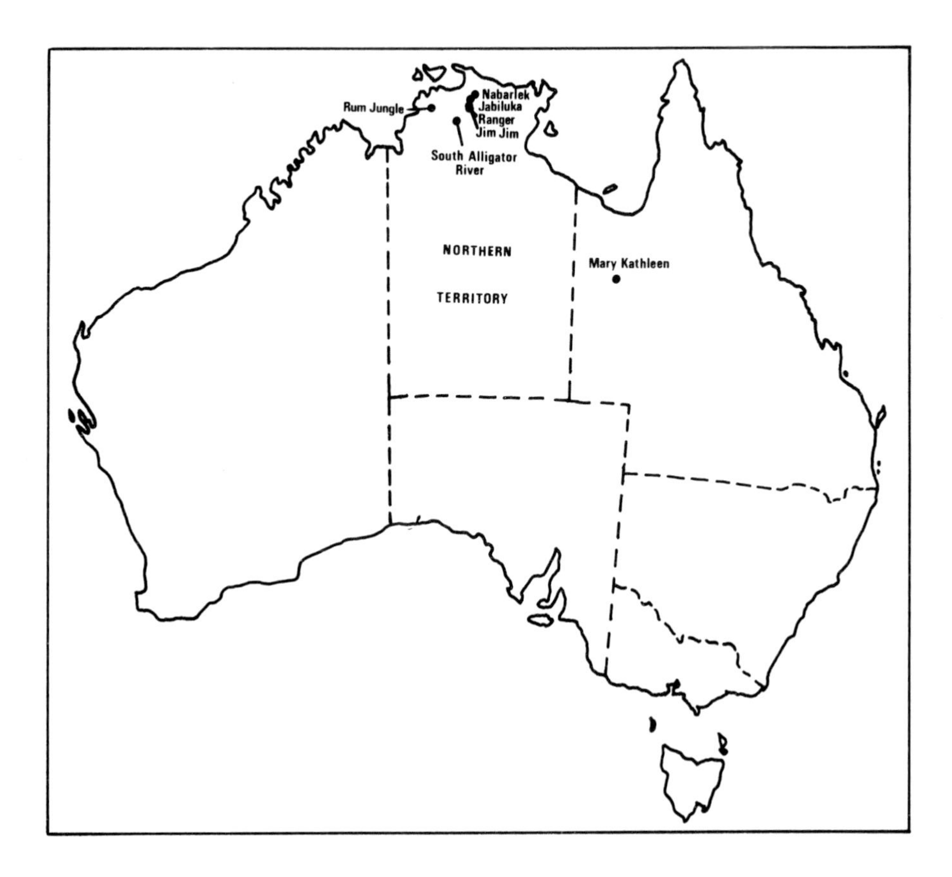

Fig. 34. Location of important hydrothermal uranium deposits in Australia.

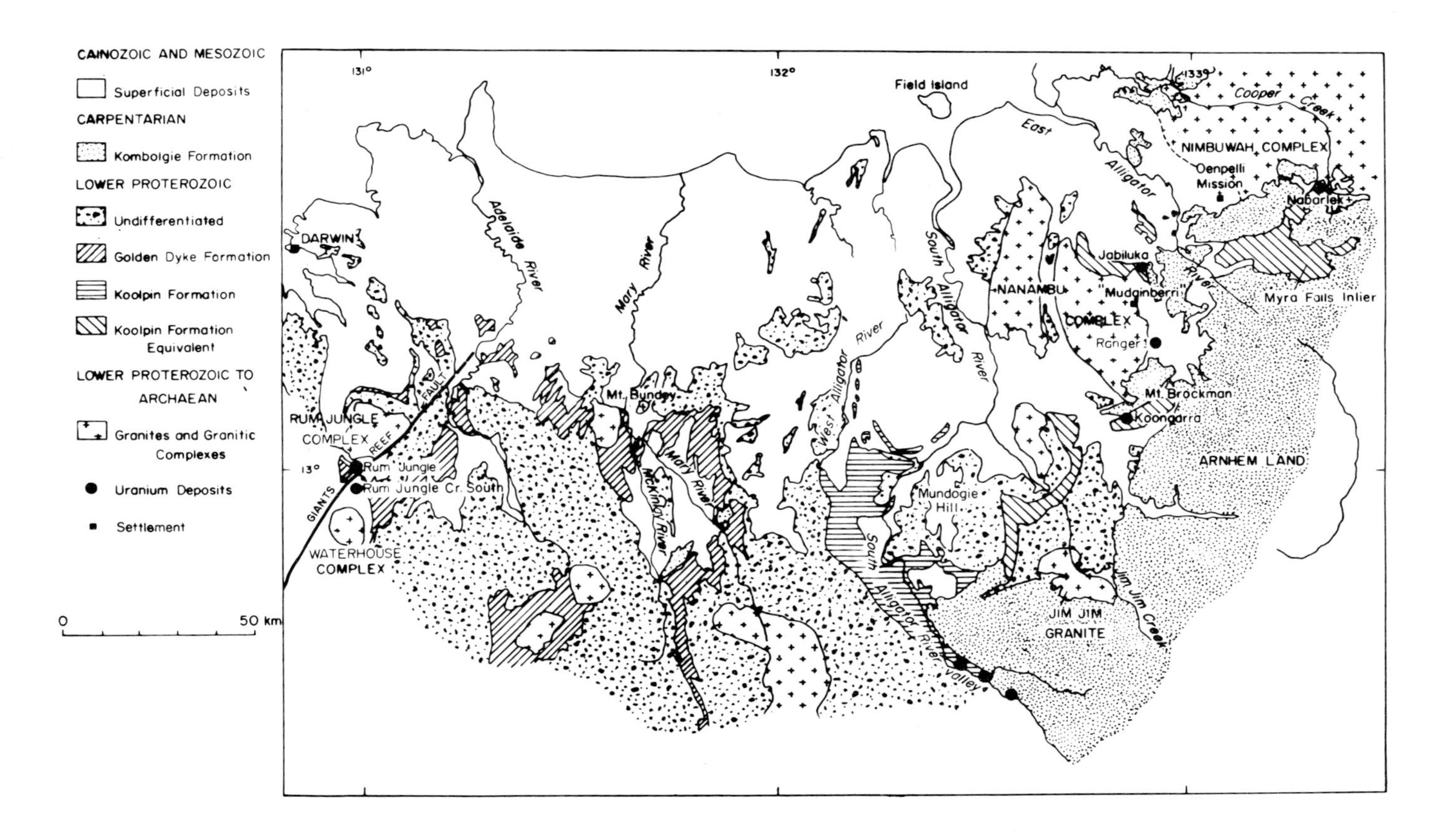

Fig. 35. Generalized geology of the Darwin region, N.T., Australia (from Dodson et al., 1974).

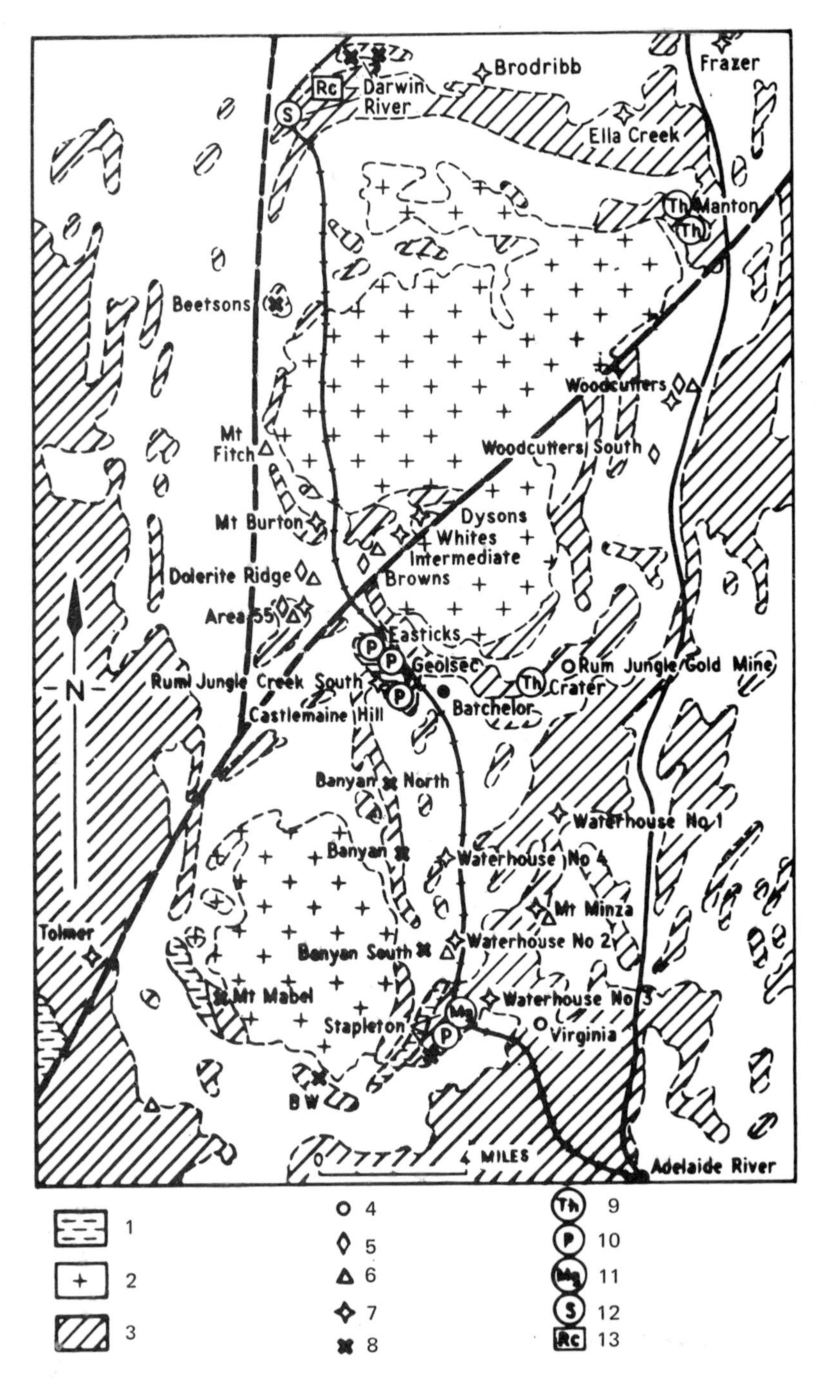

Fig. 36. Generalized geology and mineral deposits. Rum Jungle area (from Crohn, 1968). 1 = Adelaidean; 2 = granitic rocks; 3 = Lower Proterozoic; 4 = gold; 5 = silver, lead, zinc; 6 = copper; 7 = uranium; 8 = iron; 9 = thorium; 10 = phosphate; 11 = magnesite; 12 = sulphides; 13 = crushed rock.

Table 10. Mineralogy of the Rum Jungle district

aikinite (?)	hematite
anatase or rutile	libethenite
apatite	linnaeite
arsenopyrite	malachite
autunite	marcasite
azurite	meta-saleeite
bismuth	meta-torbernite
bismuthinite (?)	muscovite
bornite	native copper
bournonite	niccolite
bravoite	nontronite (?)
chalcopyrite	phosphuranylite
chalcedony	pitchblende
chlorite	pyrite
chalcocite	quartz
cornetite	rammelsbergite
covellite	sericite
cubanite	siderite
digenite	sphalerite
dihydrite	tennantite
fluorite	torbernite
galena	tourmaline
gersdorffite	wittichenite

Paragenesis:

Sequence of deposition:

1) Pyrite.
2) Pitchblende + gersdorffite(?).
3) Linnaeite + cobaltite.
4) Chalcopyrite + bornite + cubanite + digenite + covellite.
5) Development of exsolution textures in the sulfides of Stage 4.
6) Chalcopyrite showing lamellar twinning.
7) Tourmaline + quartz veins.

Wall Rock Alteration:

Late quartz and tourmaline introduced.

Ages:

Granite: 1700-1760 m.y.

Pitchblende: 650 m.y.

Ore Controls:

1. Ore occurs in folded and faulted graphitic quartz-sericite schist of low metamorphic grade.
2. Uranium mineralization occurs in north trending faults, whereas copper and lead sulfide mineralization is found in later east trending faults.

3. Secondary autuñite occurs in quartz-hematite veins which form reefs of brecciated quartz in shear zones within basement granite.

2) South Alligator River District*

Location:

Darwin region, Northern Territory, 132° 30′ E, 13° 30′ S (Figure 34).

Geology:

See Figure 37.

The South Alligator River uranium district consists of several high grade deposits distributed along a 19 km length of a major reverse fault which cuts both Lower and Upper Proterozoic rocks. The Lower Proterozoic rocks (shales, siltstones, greywackes, cherts, and basic dikes and sills) have been folded and intruded by granites. Upper Proterozoic volcanics and arenites were deposited unconformably upon these earlier units. The major uranium deposits are found in black carbonaceous and pyritic shales and siltstones of the Lower Proterozoic Koolpin Formation, and are distributed along the northeast side of a ridge of Archean basement rocks. Uranium ore bodies consist of veins, stringers and pods located in shears and cross-fractures associated with the major reverse fault zone mentioned above. Known uranium deposits occur within 100 m of the present erosion surface.

Mineralogy:

See Table 11.

Ore consists of massive and disseminated pitchblende associated with some pyrite and sparse cobalt-nickel arsenides, clausthalite and gold. Gangue consists chiefly of wall rock with some introduced quartz. Siderite is present in one deposit.

Table 11. Mineralogy of the South Alligator River district

apatite	marcasite
chalcopyrite	niccolite
clausthalite	nickel-selenides
coloradoite	pitchblende
cobalt-nickel arsenides	pyrite
galena	quartz
gersdorffite	sericite
gold	siderite
hematite	tourmaline

*Description based on Ayres and Eadington (1975), Taylor (1968), and Dodson et al. (1974), except as otherwise noted.

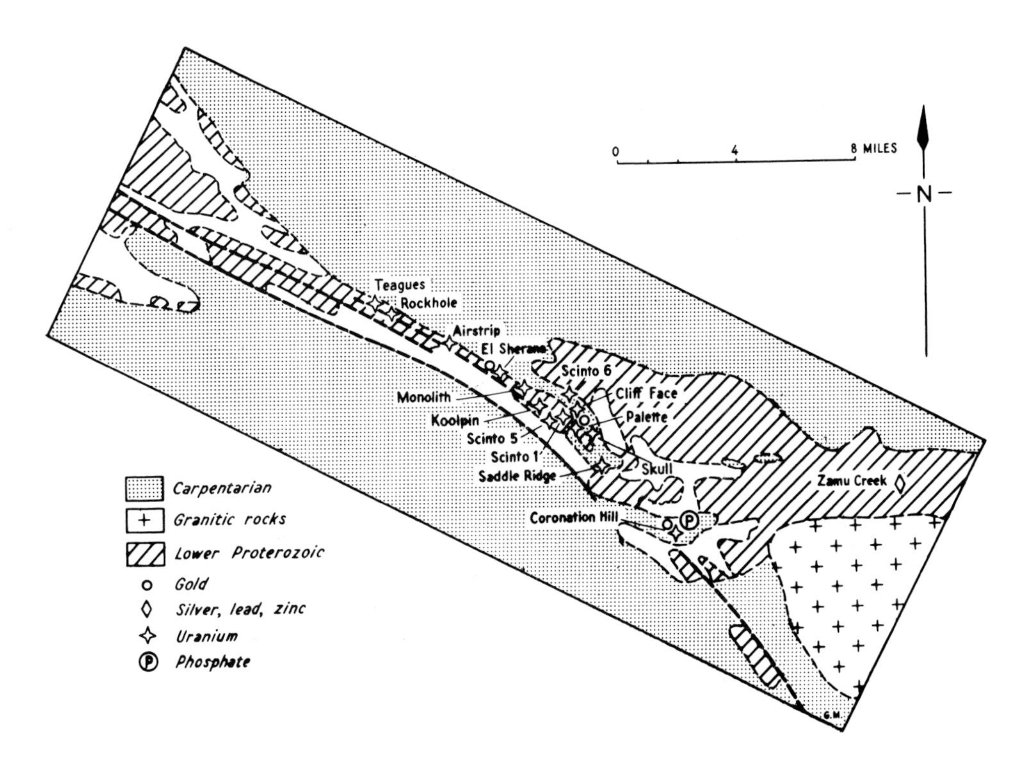

Fig. 37. Geology and mineral deposits of the South Alligator River area (from Crohn, 1968).

Paragenesis:

1. Co-Ni arsenides + pyrite.
2. Pitchblende (main).
3. Pyrite + marcasite + chalcopyrite + galena + selenides + Au + quartz + siderite + hematite.
4. Pitchblende (minor).

Wall Rock Alteration:

Kaolinization and bleaching of carbonaceous shales; sericite + tourmaline + apatite developed in wall rock.

Ages:

Pitchblende ages range from 550 to 700 m.y.; 1700-1800 m.y. old granites intrude the Lower Proterozoic sequence which hosts the ore.

Ore Controls:

1) Structural: Mineralization is localized by major faulting and associated tension fractures.
2) Lithologic: Carbonaceous shales and sandstones are favorable ore hosts, especially when adjacent to cherty ferruginous siltstones.

3) Alligator Rivers District*

Location:

Darwin region, Northern Territory, approximately 133° E and 13° S (Figure 34).

Geology:

See Figure 35.

The geology of the Alligator Rivers area is dominated by the occurrence of the Nanambu and Nimbuwah basement complexes. These are surrounded by Lower Proterozoic metasedimentary rocks, which are the metamorphosed equivalents of the sediments of the South Alligator River district to the southwest. The host rocks for the uranium deposits are locally carbonaceous quartz-chlorite-muscovite schists (Koolpin Formation equivalent). Uranium deposits in the Alligator Rivers area are shallow and often of very high grade. Important deposits of this district include Nabarlek, Jabiluka, Ranger, and Koongarra (Jim-Jim). In these deposits, pitchblende is the only important uranium mineral; it may be accompanied by chlorite, minor pyrite, copper and lead sulfides, and trace gold. Ore zones at the Ranger #1 deposit are

*Description based on Dodson et al. (1974).

chloritized. Recent data for the Jabiluka deposit indicate the existence of reserves in excess of 100,000 tons of uranium; this uranium deposit is therefore one of the largest in the world.

B) Mary Kathleen Deposit*

Location:

Queensland, east of Mt. Isa, 140° E, 20° 46′ S (Figure 34).

Geology:

See Figure 38.

The Mary Kathleen deposit occurs in a folded, faulted, and metamorphosed Lower Proterozoic sequence (the Corella Formation), consisting predominantly of quartzites, impure crystalline limestones, and siliceous and calcareous granulites, with some interbedded basic flows and sills. This sequence was intruded and contact metamorphosed by granite. The uranium orebody occupies the axial zone of a syncline, and is confined to the pyrometasomatic garnetiferous part of a thin breccia-conglomerate unit which consists of sub-angular and rounded pieces of quartzite and feldspathic rock in a matrix of fine-grained diopside and feldspar. Ore consists of uraninite, rare earth-bearing silicates, and sulfides in garnetized calc-silicate rocks.

Mineralogy:

See Table 12.

The uranium occurs in uraninite. The U/Th ratio is 5:1 and is constant throughout the ore body.

Paragenesis:

1) Allanite replaces garnet-diopsite granulite.
2) There are two generations of sulfides present; both are younger than the uraninite.
3) Uraninite replaces andradite and rarely almandite.

Table 12. Mineralogy of the Mary Kathleen deposit

allanite	galena
almandite	gummite
andradite	molybdenite
apatite	pyrite
arrojadite	pyrrhotite
beta-uranotile	rinkite
calcite	silica (amorphous)
caryocerite	stillwellite
chalcopyrite	uraninite
diopside	uranophane
fluorite	

*Description based on Matheson and Searl (1956) and Hughes and Munro (1965).

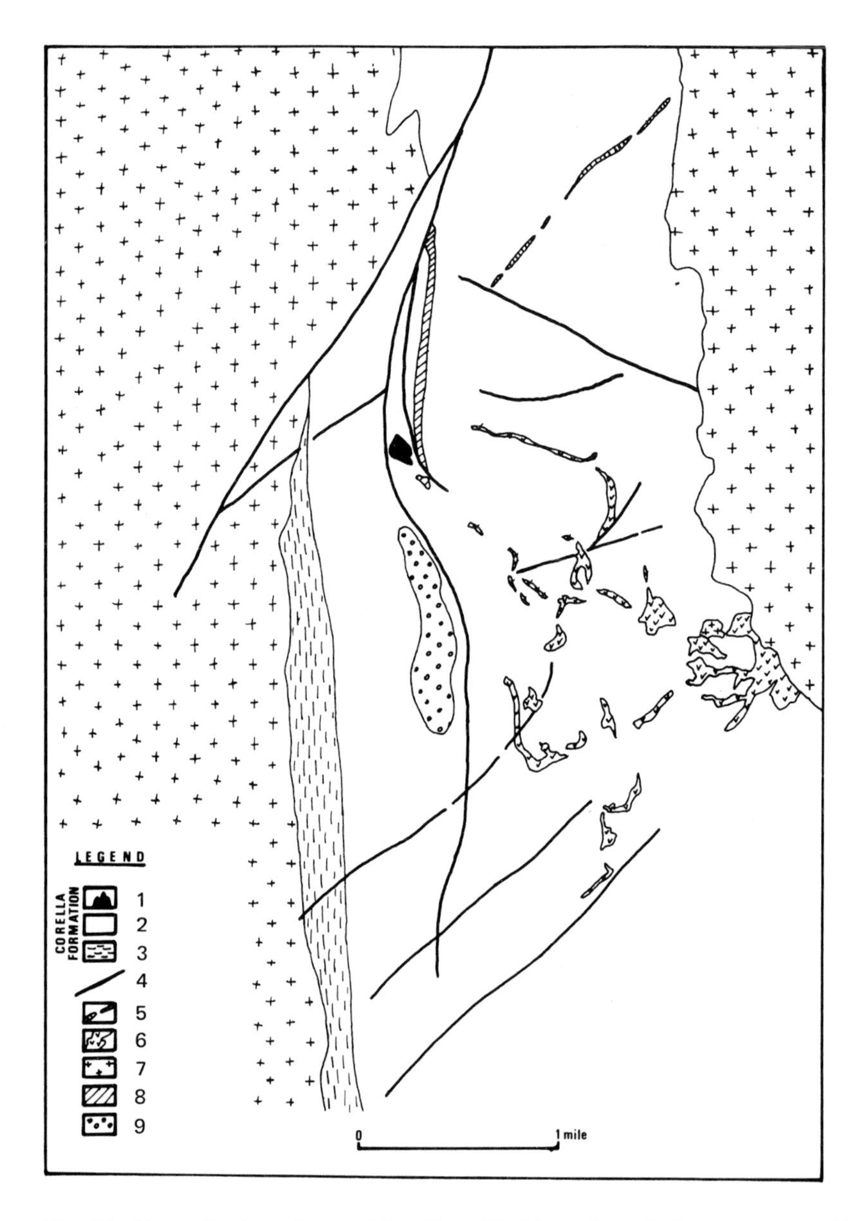

Fig. 38. Generalized geology of the Mary Kathleen deposit (redrawn from Hughes and Munro, 1965). Corella Formation: 1 = uranium orebody; 2 = calc-silicate rocks; 3 = calc-silicates and interbedded basic rocks. 4 = fault; 5 = dolerite dyke; 6 = quartz-feldspar porphyry; 7 = granite; 8 = altered diorite sill; 9 = altered basic intrusion.

Wall Rock Alteration:

1) Garnetization, especially in well jointed granulite and breccia-conglomerate.
2) Feldspathization (concurrent with uranium mineralization?).
3) Scapolitization.

Ore Controls:

1) Favorable host rocks: uranium mineralization occurs in calc-silicate rocks and especially in the garnetized zones of the breccia-conglomerate.
2) Structure: northeast, north, and northwest trending faults and accompanying joints and shears are mineralized.

Selected Australia References

Ayres, D.E. and Eadington, P.J., 1975, Uranium mineralization in the South Alligator River Valley: Mineralium Deposita, *10*, 27-41.

Crohn, P.W., 1968, The mines and mineral deposits of the Katherine-Darwin region: Australia Bur. Mineral Resources, Geology and Geophysics Bull. 82, 171-282.

Dodson, R.G., Needham, R.S., Wilkes, P.G., Page, R.W., Smart, P.G. and Watchman, A.L., 1974, Uranium mineralization in the Rum Jungle-Alligator Rivers Province, Northern Territory, Australia: in *Formation of Uranium Ore Deposits,* 551-568, Internat. Atomic Energy Agency, Vienna.

Heier, K.S. and Rhodes, J.M., 1966, Thorium, uranium and potassium concentrations in granites and gneisses of the Rum Jungle complex, Northern Territory, Australia: Econ. Geology, *61,* 563-571.

Hughes, F.E. and Munro, D.L., 1965, Uranium ore deposit at Mary Kathleen: in McAndrew, J., ed., *Geology of Australian Ore Deposits,* 8th Commonwealth Min. Metall. Cong., *1*, 256-263.

Langford, F.F., 1974, A supergene origin for vein-type uranium ores in the light of the Western Australian calcrete-carnotite deposits: Econ. Geology, *69*, 516-526.

Matheson, R.S. and Searl, A.R., 1956, Mary Kathleen uranium deposit, Mount Isa-Cloncurry district, Queensland, Australia: Econ. Geology, *51*, 528-540.

Roberts, W.M.B., 1960, Mineralogy and genesis of White's ore body, Rum Jungle uranium field, Australia: Neues Jahrb. Mineralogie Abh., *94,* 868-889.

Spratt, R.N., 1965, Uranium ore deposits of Rum Jungle: in McAndrew, J., ed., *Geology of Australian Ore Deposits,* 8th Commonwealth Min. Metall. Cong., *1*, 201-206.

Taylor, J., 1968, Origin and controls of uranium mineralization in the South Alligator Valley: Symposium on Uranium in Australia, Australasian Inst. Mining and Metallurgy, Rum Jungle Branch, 32-44.

Walpole, B.P., Dunn, P.R., Crohn, P.W. and Randal, M.A., 1968, Geology of the Katherine-Darwin region, Northern Territory: Australia Bur. Mineral Resources, Geology and Geophysics Bull. 82, 304 pp.

Whittle, A.W.G., 1960, Contact mineralization phenomena at the Mary Kathleen uranium deposit: Neues Jahrb. Mineralogie Abh., *94*, 798-830.

III) EUROPE

The distribution of various types of hydrothermal uranium deposits in Europe is shown in Figure 39. For more detail, the reader is referred to the Carte Metallogénique de l'Europe, 1:2,500,000, 1st edition, 1968-1970, published by the United Nations. This map clearly shows the spatial relation between uranium deposits and Hercynian granitic massifs.

A) CENTRAL EUROPE

1) Erzgebirge (Krusné Hory) Region *

The Erzgebirge is located in northwestern Bohemia, straddling the border between Czechoslovakia and East Germany (Figure 40). A generalized view of the regional geology and the locations of major ore deposits are given in Figure 41. The country rocks of the Erzgebirge consist of a Proterozoic and early Paleozoic metamorphic sequence which has been intruded by numerous Hercynian granitic stocks. Ruzicka (1971) has reviewed the geology and ore deposits of the Erzgebirge. The maps of Tischendorf et al. (1965) and Chrt et al. (1968) clearly illustrate the relation of the ore deposits to the regional geology. The principal uranium-bearing ore districts of the region are Jáchymov (St. Joachimsthal), Johanngeorgenstadt, and Horni Slavkov. Less important pitchblende occurrences include Schneeberg, Marienberg, and Freiberg.

In the Erzgebirge region, uranium veins occupy fractures, but the deposition of pitchblende within the veins seems to have been controlled by host rock lithology. Most pitchblende was deposited where veins intersect chloritized and pyritized biotite gneiss, amphibolite, skarn, and other mafic or graphitic rock units. Granites are very poor host rocks for uranium veins in the Erzgebirge; uranium veins are found almost exclusively in the metamorphic country rocks.

Pitchblende is the only important hydrothermal uranium mineral of the Erzgebirge region. Carbonates are the predominant gangue minerals of the uranium stage. The formation of the mineral deposit types of the Erzgebirge region follows the sequence:

1) Magnetite in skarns.
2) Tin-tungsten mineralization in greisens.
3) Polymetallic veins (including uranium).
4) Barite-fluorite veins.

*Description based on Ruzicka (1971) and Schneiderhöhn (1941).

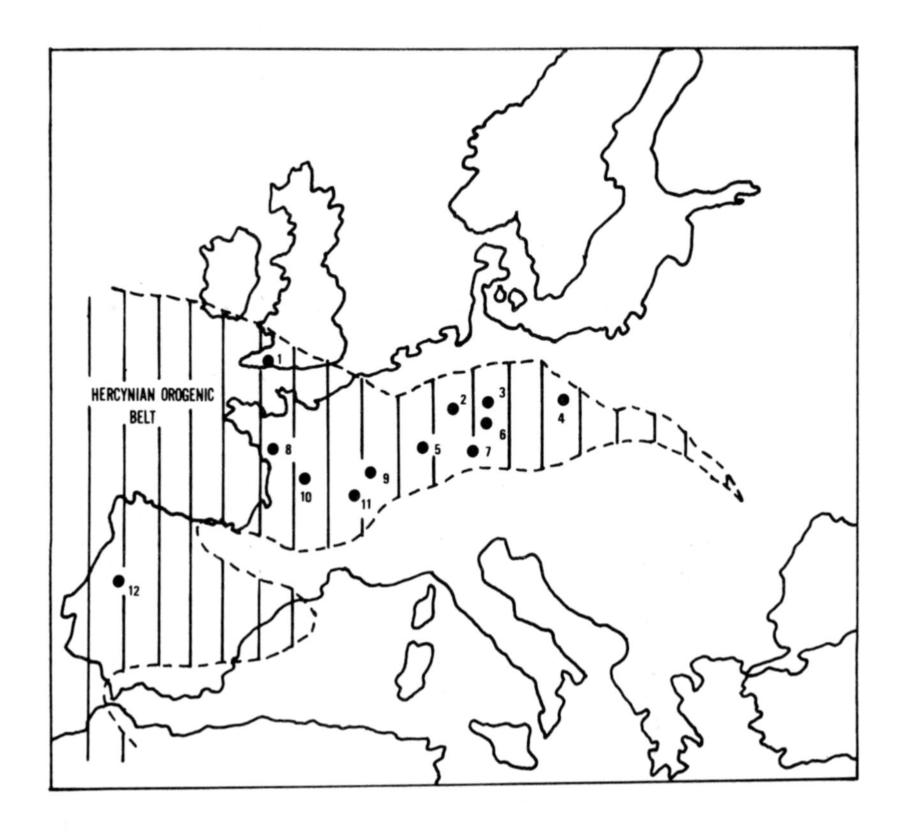

Fig. 39. Location of hydrothermal uranium districts in Europe; (1) Cornwall, (2) Thuringia, (3) Saxony and Bohemia (Jáchymov, Schneeberg, etc.), (4) Schmiedeberg, (5) Wittichen, (6) Dürrinaul, (7) Wölsendorf, (8) Vendée, (9) North-east Massif Central, (10) Limousin, (11) Forez, (12) Beira (modified from Geffroy and Sarcia, 1955; Centre d'Études Nucléaires de Saclay).

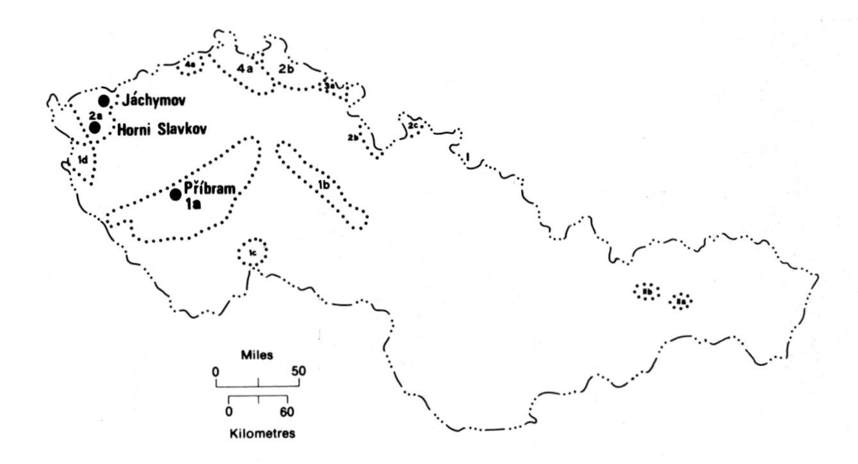

Fig. 40. Uranium-bearing regions in Czechoslovakia (modified from Ruzicka, 1971). Dotted line = boundary of uranium-bearing region. 1a = Central Bohemian Pluton uranium-bearing region; 1b = Labe Lineament uranium-bearing region; 1c = Central Moldanubian Pluton uranium-bearing region; 1d = Western Bohemian uranium-bearing region; 2a = Karlovy Vary Massif uranium-bearing region; 2b = West Sudetes uranium-bearing region; 2c = Rychlebské hory uranium-bearing region; 3a = Intra-Sudetic Basin uranium-bearing region; 4a = Cretaceous sediments uranium-bearing region of North Bohemia; IIa = Spiš — Gemer Ore-bearing Mountains uranium-bearing region; IIb = Low Tatra Mountains uranium-bearing region.

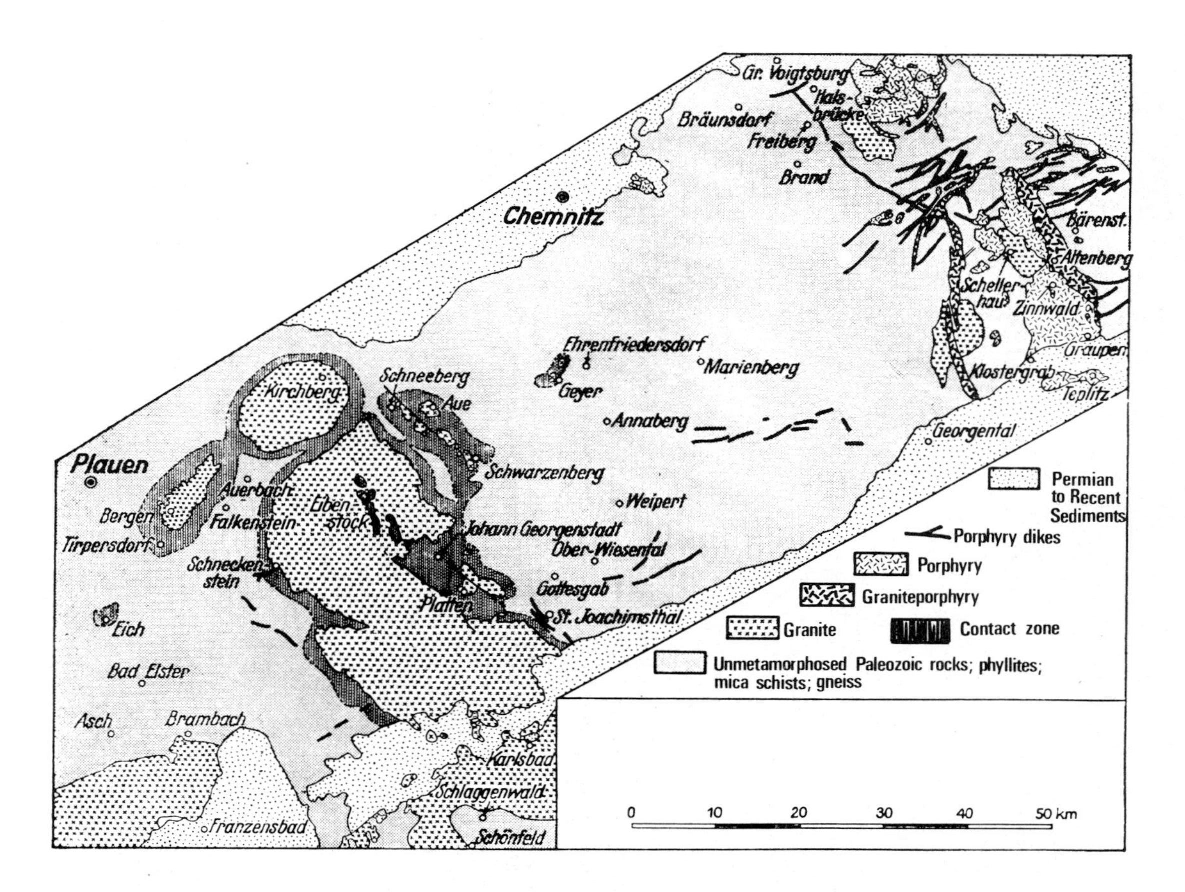

Fig. 41. Generalized geology of the Erzgebirge (modified from Schneiderhohn, 1941).

a) Jáchymov (St. Joachimsthal) District*

Location:

Southwest Erzgebirge, northwest Czechoslovakia, 50° 23′ N, 12° 54′ E (Figure 40).

Geology:

See Figure 41.

The Jáchymov district covers an area of about 35 km^2 and contains approximately 200 ore veins and splits. The country rocks of the district consist of a metamorphic complex (Proterozoic and early Paleozoic mica schists, phyllites, gneisses, calc-silicates, and amphibolites) which has been intruded by Hercynian granitic bodies. Two major types of granites occur in the Jáchymov district: an earlier "normal granite" and a later "autometamorphic granite". The normal granite shows no discernible post-magmatic alteration effects, whereas the autometamorphic granite has undergone albitization, muscovitization, and greisenization. Tin mineralization is associated with the greisenized granites. Hydrothermal mineralization followed the emplacement of, and is spatially related to, the autometamorphic granite bodies. The sequence of hydrothermal vein mineralization stages is as follows:

1. Early sulfides.
2. Quartz.
3. Pitchblende
4. Arsenides with silver.
5. Arsenides with bismuth.
6. Sulfarsenides.
7. Late sulfides.

Pitchblende mineralization (Stage 3) is the most abundant type at Jáchymov. There are four major fault (and vein) systems in the district;these trend north, northwest, east, and northeast. Only the north trending system contains important pitchblende veins. Some uranium veins contain only pitchblende and dolomite. Pyritic and graphitic muscovite-biotite schists are the most common ore host rocks for ore at Jáchymov.

Mineralogy and Paragenesis:

The paragenesis is outlined in Figure 42. In addition to the minerals listed in this figure, anhydrite, gypsum, marcasite, and adularia are found in the veins at Jáchymov.

*Description based on Ruzicka (1971), except as otherwise noted.

Minerals \ Mineralization stage	Early sulphides	Quartz	Pitchblende	Arsenides with native Ag	Arsenides with native Bi	Sulphoarsenides	Later sulphides
Quartz	●	● ● ●		●	●		
Arsenopyrite	●						●
Pyrite	●	●	●				●
Galena	●						●
Sphalerite	●						●
Chalcopyrite	●						●
Bornite	●						
Ankerite		●					
Hematite		●					
Dolomite			●			●	
Pitchblende			●	● ● ●			
Fluorite			●				
Native silver				● ●			
Native bismuth					●		
Skutterudite				● ●	● ●		
Rammelsbergite				● ●	●		
Niccolite				● ●	●		
Safflorite				●	●		
Löllingite				●	●	●	
Gersdorffite				●			
Bismuthinite					●		
Native arsenic						●	
Proustite						●	
Pyrargyrite						●	
Argentite						●	
Sternbergite						●	
Stephanite						●	
Antimonite						●	
Realgar							●
Tennantite							●
Calcite							●

Fig. 42. Paragenesis of the Jáchymov district (from Bernard et al., 1968).

Notes: (1) The fluorite is dark purple and has a fetid odor when crushed ("Stinkspat").
(2) Dolomite is the major gangue mineral associated with pitchblende; it is reddish in the vicinity of pitchblende.
(3) Vein quartz is typically chalcedonic or smoky.
(4) There are several minor generations of pitchblende following Stage 3.
(5) Hematite is not a common vein mineral, but red coloration is frequently associated with pitchblende.

Zoning:

Zonation of vein type and mineralogy are a function both of depth and distance from the granite contact. In general, the early stages of mineralization are deeper and closer to the granite contact, and the later stages are shallower and farther from the granite intrusive. The generalized zoning pattern is as follows:

Upper zone: Arsenides + silver.
Middle zone: Arsenides + bismuth.
Lower zone: Pitchblende.

Quartz is the dominant gangue mineral at depth, whereas carbonates are more important at shallower levels. The richest uranium mineralization occurs in veins at a distance of 150 to 400 m from the granite contact. Pavlů (1972) states that in veins of the sulfarsenide stage the quantity of sulfantimonides increases with depth; he also points out that in veins of the arsenide stages nickel predominates in the upper levels and cobalt is more important at depth.

Wall Rock Alteration:

	Rock type	
Stage	Metasediments	Granites
Pre-ore:	Pyritization, chloritization, and graphitization along shears.	Greisenization, albitization, and silicification.
Syn-ore:	Silicification, carbonatization, and hematization	Sericitization.
Post-ore:	Sericitization, silicification, and kaolinization.	Sericitization and kaolinization.

Ages:

Normal granite: 320 m.y.
Automectamorphic granite: 260-280 m.y.
Main stage of pitchblende deposition: 220-230 m.y.
Minor pitchblende stages: 5-160 m.y.

Ore Controls:

Structural:

1) North trending faults related to a large anticlinal structure.
2) Pitchblende lenses are localized by vein intersections and by changes in vein strike, dip, and thickness.

Lithologic:

Scapolitized rocks and metasedimentary rocks containing sulfides, graphite, and abundant biotite and amphibole are favorable host rocks. Unfavorable rock types include granite, quartz-sericite gneiss, and muscovite gneiss.

b) Horni Slavkov (Schlaggenwald) District*

Location:

Southwest Erzgebirge, northwest Czechoslovakia, southwest of Jáchymov, 50° 08′ N, 12° 50′ E (Figure 40).

Geology:

Proterozoic and Lower Paleozoic metasedimentary rocks are intruded by Hercynian granites. As at Jáchymov, both "normal" and "autometamorphic" granites are present. The autometamorphic granites have been albitized, kaolinized, sericitized and greisenized. Tin-tungsten mineralization accompanied greisenization of the autometamorphic granites, and hydrothermal vein mineralization followed. The metasediments consist mostly of high grade biotite gneisses. Locally these gneisses contain beds of amphibolite, amphibole gneiss, and skarn.

There are several vein mineralization stages at Horni Slavkov, but veins of the pitchblende-dolomite stage are the most abundant. In general, quartz is more abundant than carbonate in the veins of the district, but quartz-rich stages contain less pitchblende.

Mineralogy:

Pitchblende in veins is accompanied by dolomite, ankerite, calcite, siderite, fluorite and pyrite, all of which are less abundant and more irregularly distributed than pitchblende. Fluorite is usually dark purple in the vicinity of pitchblende.

Wall Rock Alteration:

Hematization is the predominant alteration both of vein carbonates and of the host rocks.

*Description based on Ruzicka (1971).

Ore Controls:

Favorable host rocks for uranium mineralization are amphibolite interlayered with paragneiss, amphibole gneiss, and biotite gneiss. Orthogneiss is unfavorable, and uranium mineralization is practically absent where veins cut this rock type. Quartzite, biotite-muscovite gneiss, aplitic gneiss, and granite are also unfavorable host rocks. Ore is localized where veins change thickness, dip, or strike, and where veins cut metasediments that contain pyrite, biotite, amphibole, or graphite.

c) Johanngeorgenstadt District*

Location:

Southwest Erzgebirge, East Germany, northwest of Jáchymov, 50° 26′ N, 12° 44′ E (Figure 41).

Geology:

See Figure 41.

Host rocks are mostly phyllites, with some quartzites and amphibolites. These metamorphic rocks have been intruded by granitic bodies. Hematite occurs as an accessory mineral in the phyllites, quartzites, granites, and porphyritic microgranites. The amphibolites contain magnetite and hematite.

Mineralogy and Paragenesis:

Older veins:

1) Tin veins, containing quartz + cassiterite + tourmaline (with apatite, topaz, fluorite, nacrite, hematite, arsenopyrite, chalcopyrite, pyrite, sphalerite, wolframite and molybdenite).
2) Quartz + sphalerite + galena veins, containing minor pyrite, chalcopyrite and arsenopyrite.

Younger veins:

3) Cobalt-silver veins contain quartz (partly pseudomorphic after barite and calcite), dolomite, calcite, native silver, argentite, proustite, marcasite, pyrite, galena, skutterudite, chloanthite, sternbergite, native bismuth, bismuthinite, pitchblende, niccolite, millerite, galena, sphalerite, chalcopyrite, and arsenopyrite; pitchblende is commonly associated with ankerite, and sometimes with galena, chalcopyrite and native bismuth. Among the cobalt-silver veins the sequence of deposition is cobalt-nickel ores with bismuth,

*Description based on Viebig (1905).

followed by silver ores with bismuthinite and marcasite.

4) Iron-manganese veins contain mostly quartz, hematite, pyrolusite, polianite, psilomelane and manganite, with some chalcopyrite; quartz replaces calcite, barite, anhydrite and fluorite.

Zoning:

1) Tin and iron-manganese veins are mainly restricted to the granitic intrusives.
2) The pitchblende content of veins in schist increases with depth.
3) Silver-bismuth (and uranium?) veins become barren upon entering granite.

Ore Controls:

The major bismuth-silver-uranium ore shoots occur in a restricted vertical interval which coincides with a zone of thin-bedded, fine-grained pyritic schists. Rich ores generally end within a few meters above this favorable horizon, but they may extend as much as 80-100 m below the favorable horizon.

d) Schneeberg District*

Location:

Erzgebirge region, East Germany, northwest of Jáchymov, 50° 36′ N, 12° 39′ E (Figure 41).

Geology:

See Figure 41.

Country rocks consist of Proterozoic and Lower Paleozoic schists cut by granitic intrusives. The vein district is 5 × 3 km.

Mineralogy and Paragenesis:

Sequence of vein formation:

1. Tin-tungsten veins.
2. Tin-copper veins.
3. Copper veins.
4. Siliceous lead veins.
5. Siliceous cobalt-bismuth veins.
6. Baritic cobalt-silver veins.
7. Hematite veins.

Baritic cobalt-silver veins contain pitchblende, barite, fluorite, ankerite, calcite, quartz, cobalt-nickel arsenides, native bismuth, native silver, argentite and galena.

*Description based on Schneiderhöhn (1941), except as otherwise noted.

Zoning:

Viebig (1905) states that veins become lean or barren upon entering granite from schist. The amount of uranium present increases with depth.

e) Freiberg District*

Location:

Northeastern Erzgebirge region, East Germany, 50° 55′ N, 13° 21′ E (Figure 41).

Geology:

See Figure 43.

The host rocks are gneiss, schist and phyllite cut by granitic stocks. The ore district measures approximately 15 × 20 km.

Mineralogy and Paragenesis:

See Figure 44.

Generalized sequence of vein formation:

1. Pyritic lead veins.
2. Ankeritic veins containing pitchblende.
3. Siliceous veins.
4. Fluorite-barite-lead veins.

The paragenetic sequence within pitchblende-bearing Stage II veins:

1. Galena + quartz + silver minerals.
2. Galena + arsenopyrite + sphalerite.
3. Quartz + reddish barite + fluorite + chalcedony + amethyst + chalcopyrite + hematite + pitchblende I (sulfides are very sparse during this sub-stage).
4. Galena + reddish carbonates.
5. Fluorite + barite + galena.
6. Bismuth, cobalt, nickel and silver minerals + pitchblende II.

Ore Controls:

Veins are richest in grey gneiss, barren in red gneiss, and pinch out in quartz porphyry. The grey gneiss contains biotite which is strongly altered close to the veins. The red gneiss contains only muscovite. Ore shoots are particularly rich beneath garnet-mica schists.

*Description based on Bernard et al. (1968), Leutwein (1957), and Schneiderhöhn (1941).

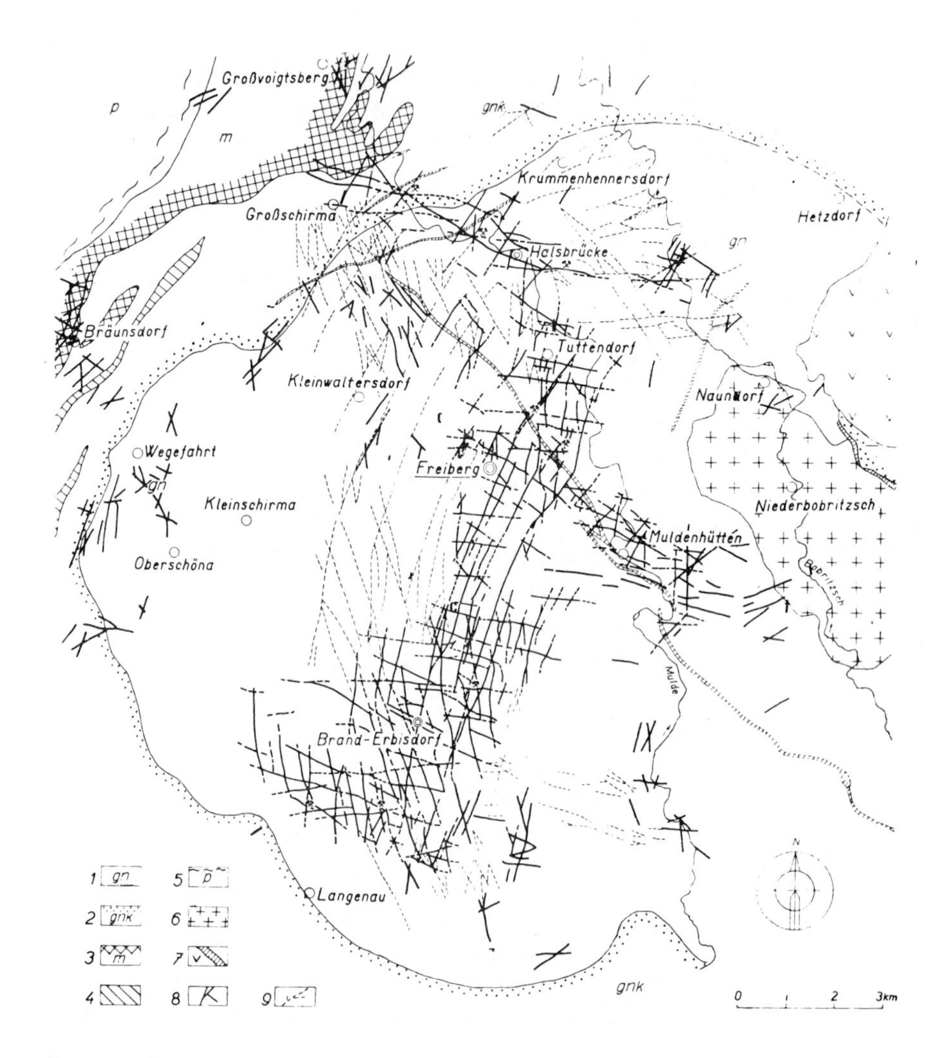

Fig. 43. Geology of the Freiberg district (from Bernard et al., 1968). 1 = gneiss; 2 = Marienberg gneiss; 3 = mica schist; 4 = red gneiss; 5 = phyllite; 6 = granite; 7 = porphyry; 8 = ore veins; 9 = barren faults.

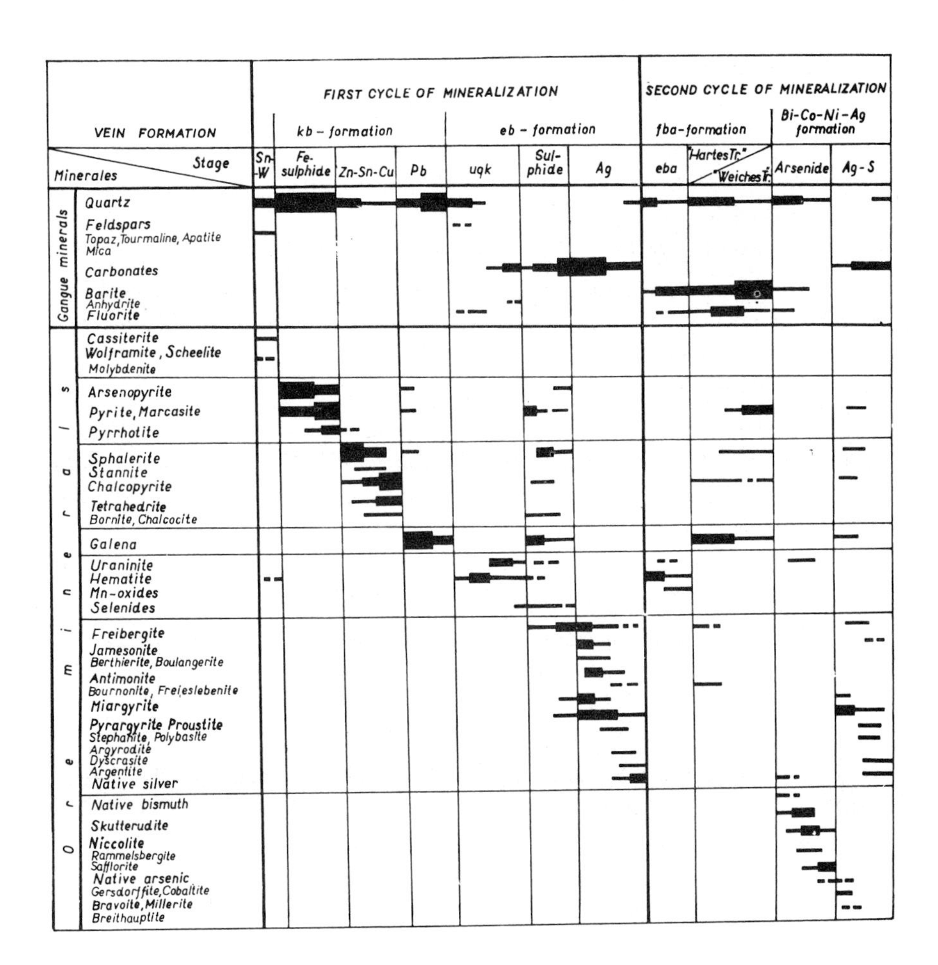

Fig. 44. Paragenesis of the Freiberg district (from Bernard et al., 1968).

2) Príbram District*

Location:

Central Bohemian Pluton, western Czechoslovakia, 49° 40′ N, 13° 14′ E (Figure 40).

Geology:

See Figure 45.

Country rocks consist of Proterozoic and Cambrian pelitic sediments and conglomerates which have been intruded by granitic rocks. Uranium mineralization extends to a depth of 1500 m in the Príbram district.

Mineralogy:

Vein minerals include ankerite, arsenopyrite, calcite, dolomite, galena, nickel-cobalt minerals, pitchblende, pyrite, quartz, siderite, sphalerite and uranoan-anthraxolite.

Paragenesis:

Generalized sequence of deposition:

1. Siderite + quartz + sphalerite + nickel-cobalt minerals + galena + dolomite + arsenopyrite.
2. Calcite + dolomite + ankerite + pitchblende.
3. Calcite.

At the Vrancice deposit (Figure 45) the following detailed sequence is observed:

1. a) Arsenopyrite + pyrite + quartz + sphalerite + pyrrhotite.
   b) Ankerite + jamesonite + galena.
   c) Calcite.
2. a) Quartz + hematite.
   b) Nickel minerals + sphalerite + siderite.
   c) Quartz + bournonite + tetrahedrite.
   d) Barite + galena + chalcopyrite + bornite + chalcocite.
   e) Quartz + willemite + calcite + native silver + uraninite.
   f) calcite + goethite.

Wall Rock Alteration:

Chloritization, sericitization, hematization and pyritization are reported. Hematization is most common in association with pitchblende-carbonate veins.

*Description based on Ruzicka (1971), Bernard et al. (1968), Písa (1966), and Pluskal (1970).

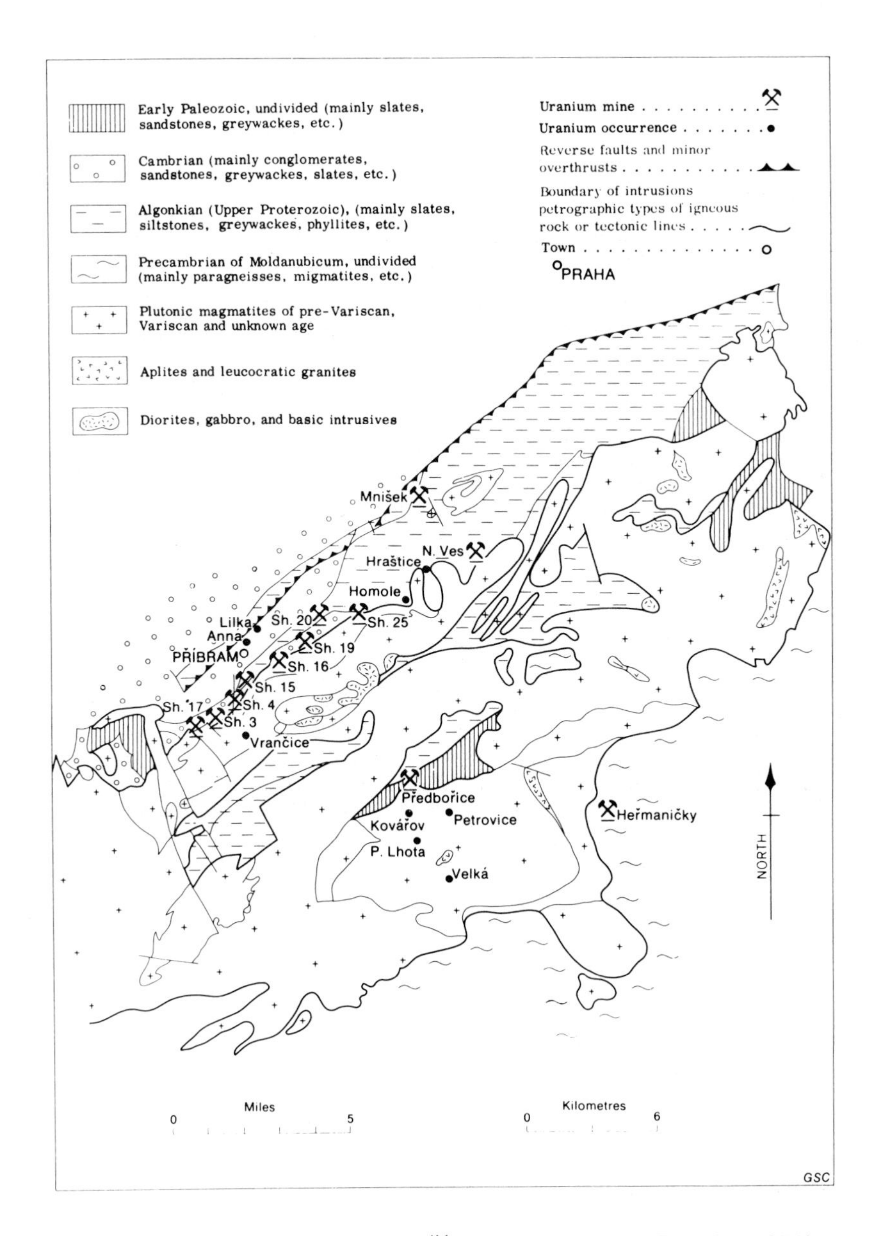

Fig. 45. Generalized geology of the Příbram district (from Ruzicka, 1971).

Ages:

The intrusives of the Príbram district are dated as Lower Carboniferous; the age of the main pitchblende stage is about 270 m.y.

Ore Controls:

1) Favorable ore structures include anticlinal folds, faults, dikes and sills, and fissures that ramify or are intersected by other veins.
2) Favorable host rocks include pyritic or graphitic shales, siltstones and greywackes. Uranium is often found in graphitic shear zones.
3) A correlation exists between vein thickness and the amount of uranium mineralization: the thicker the vein, the more pitchblende is present.

3) Labe Lineament Region*

Location:

West-central Czechoslovakia, 50° N, 15° 30′ E to 49° 30′ N, 16° 20′ E (Figure 40).

Geology:

Host rocks are graphitic and pyritic paragneisses, intercalated with amphibolites, quartzites and metamorphosed limestones.

Mineralogy and paragenesis:

Minerals present include berzelianite, bornite, calcite, chalcocite, chalcopyrite, chlorite, clausthalite, eucairite, graphite, hematite, pitchblende, pyrite, quartz, and umangite.

Depositional sequence:

1. Quartz + hematite or carbonate + hematite with sulfides.
2. Carbonate + pitchblende + graphite + hematite + chlorite + some metallic minerals.
3. Quartz + hematite + carbonates + pyrite.

Zoning:

The youngest stage of mineralization is usually developed at deposit margins.

Wall Rock Alteration:

Chloritization and carbonatization.

*Description based on Ruzicka (1971).

Ore Controls:

Shear zones and fissures related to the Labe Lineament; flexures, intersections and broadening of shear zones and fissures. The main shear zones carry disseminated uranium mineralization within the graphitic, chloritic, and pyritic material of the brecciated and mylonitized wall rocks.

B) WESTERN EUROPE

1) France*

General Statement:

There are four major regions in France containing hydrothermal uranium deposits: Morvan, Limousin, Forez, and Vendée; these regions are located within Hercynian massifs, three occurring in the Massif Central (Figure 46). All uranium deposits are found in or, in the case of the Vendée region, immediately adjacent to intrusive bodies of two-mica granite. These granites have an anomalously high uranium content (15-20 ppm), exhibit a tendency toward sodium enrichment, and are frequently accompanied by an aureole of tin-tungsten veins. In the two-mica granites 50 to 70% of the contained uranium occurs as microscopically disseminated uraninite. This uraninite is low in thorium, and its presence seems to coincide with the deuteric transformation of biotite to muscovite. In clear contrast to the two-mica granites, 70% of the uranium contained in biotite granites occurs in "refractory" accessory minerals such as zircon, monazite, and apatite. Two-mica granites have ages of about 300 m.y.; the age of contained uranium mineralization averages 250 m.y.

Hypogene uranium occurs as pitchblende and minor coffinite. The uranium is accompanied by a characteristically sparse paragenesis, chiefly pyrite, marcasite, galena, sphalerite and chalcopyrite. Selenides are present in some deposits. In general, the uranium deposits are shallow, occurring within a few hundred meters of the surface. Ore bodies consist of uranium-bearing, silicified fault and fracture zones, stockworks of small veins, and mineralized zones within episyenite (alkali metasomatized, carbonatized, and desilicified granite). Uranium mineralization is especially intense where veins intersect lamprophyre dikes.

*Description based on Gangloff (1970), Geffroy (1971) and Sarcia et al. (1958).

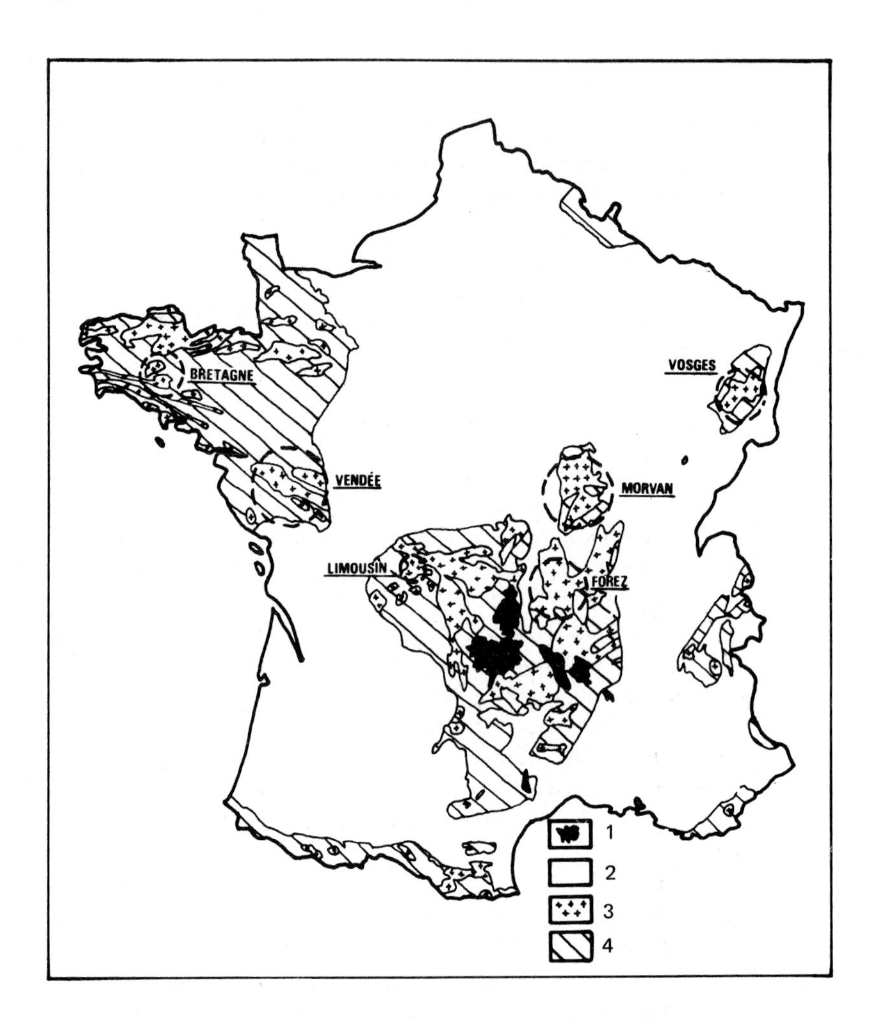

Fig. 46. Hydrothermal uranium districts in France (redrawn from Geffroy and Sarcia, 1960). 1 = Tertiary volcanics; 2 = Post-Paleozoic sedimentary rocks; 3 = granite; 4 = metamorphic and Paleozoic sedimentary rocks.

The most important type of hydrothermal uranium deposit occurring in France is "le type uranifère propre". Uranium deposits of this type occur as veins within granite and are characterized by the occurrence of pitchblende in a predominantly siliceous gangue, accompanied by minor fluorite, carbonate, and barite, some coffinite, and small amounts of sulfides. A few French uranium deposits (e.g. Bauzot in the Morvan region) belong to the "fluorite type", in which sulfides are more abundant and fluorite is an important gangue mineral. There are no representatives of the "Co-Ni-Ag-As-Bi type" among the important uranium deposits of France. The diagrams presented in Table 13 summarize the important features of the mineralogy and paragenesis of several French hydrothermal uranium deposits; additional paragenetic diagrams are included with the descriptions of deposits in the several regions.

a) Limousin Region*

Location:

Northwest Massif Central, France 45° 45′ N, 1° 15′ E (Figure 46).

Geology:

See Figure 47.

The uranium deposits of the Limousin region occur in alkaline two-mica granites which have been cut by lamprophyre and aplite dikes. Ore bodies are found in silicified fault breccia zones, as large mineralized bodies within episyenite, as stockworks of thin veins associated with secondary structural features, and where fissure veins intersect lamprophyre dikes. Most mineralized zones are marked by some degree of hematization of the host rock. The hypogene vein mineralogy is simple. Pitchblende is the dominant vein mineral, occurring with small amounts of pyrite and marcasite, and rare galena, chalcopyrite, and bismuthinite. Gangue is not abundant and may be absent; when gangue minerals are present, carbonate or chalcedony are dominant. In addition, fluorite and sometimes barite are present.

Mineralogy and Paragenesis:

See Tables 13 and 14 and Figures 48a-b.

Typical sequence of events for the formation of uranium veins in granite:

1. Compression, shearing, and pitchblende deposition.
2. Relaxation of stress, accompanied by deposition of pitchblende + silica + fluorite + pyrite + marcasite.

*Description based on Lenoble and Gangloff (1958), Sarcia (1958), and Sarcia et al. (1958).

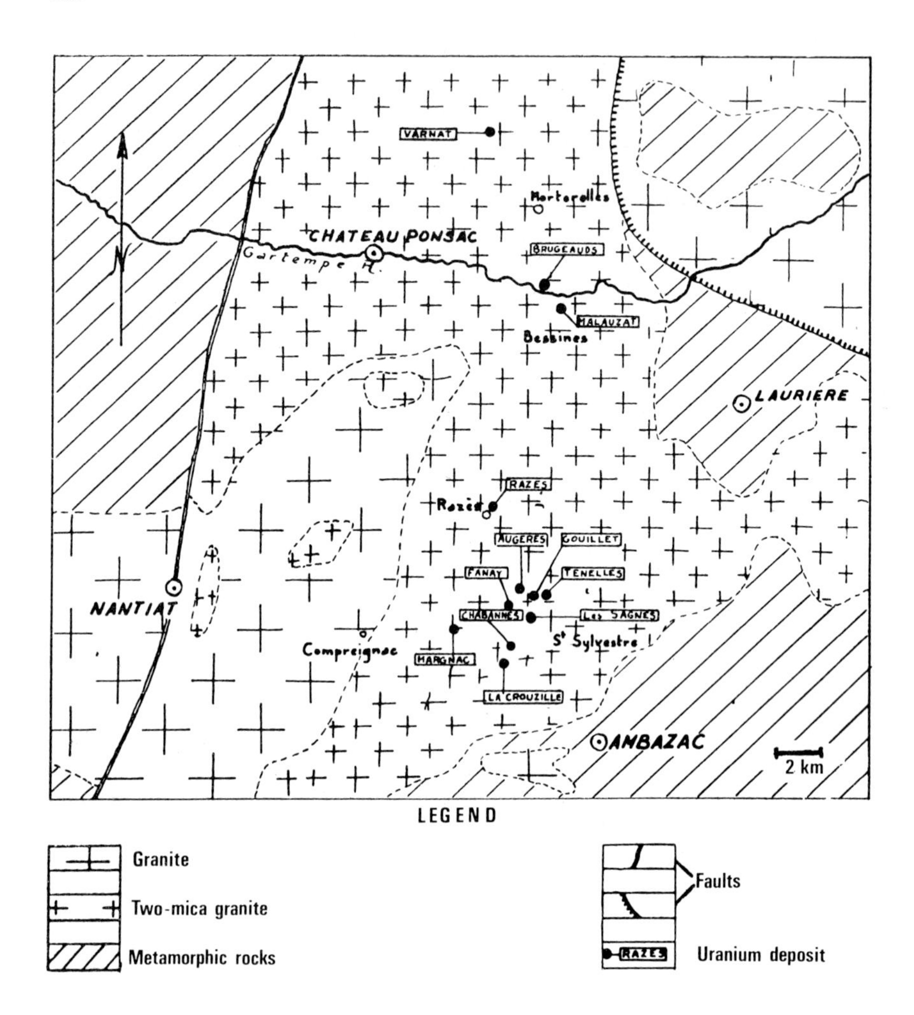

Fig. 47. Geology of the Limousin region (modified from Geffroy and Sarcia, 1955; Centre d'Études Nucléaires de Saclay).

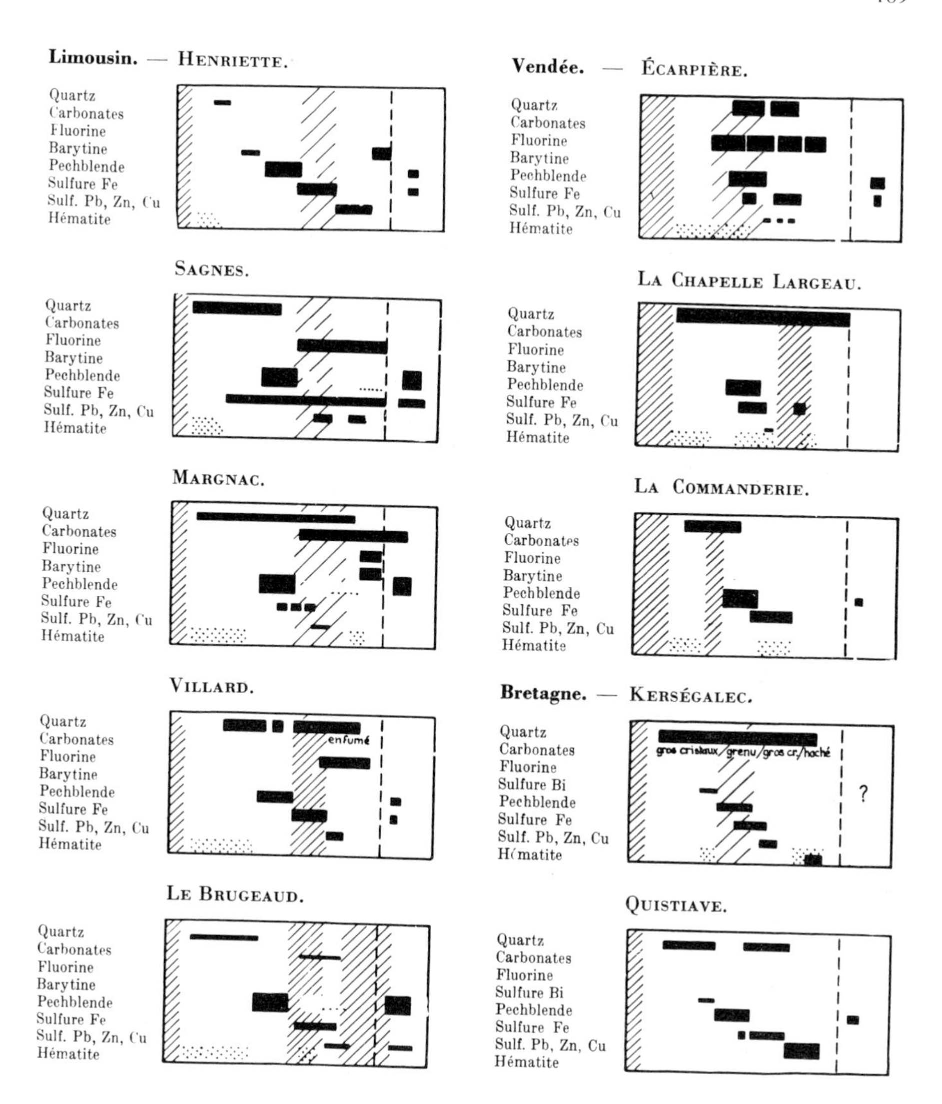

Table 13. Paragenetic diagrams for some hydrothermal uranium deposits in France; supergene deposition is shown to the right of the vertical dashed line in each diagram modified from Geffroy and Sarcia, 1960; Centre d'Études Nucléaires de Saclay).

Table 13 continued.

Forez. — BIGAY.

Quartz
Carbonates
Fluorine
Barytine
Pechblende
Sulfure Fe
Sulf. Pb, Zn, Cu
Hématite

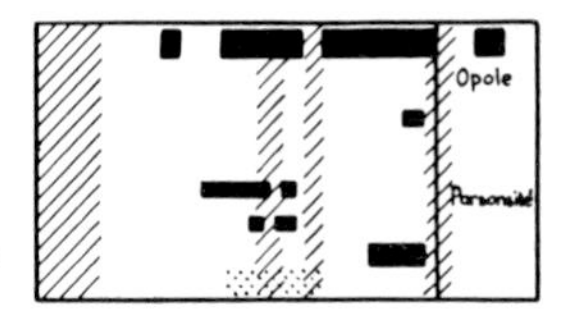

LIMOUZAT.

Quartz
Carbonates
Fluorine
Barytine
Pechblende
Sulfure Fe
Sulf. Pb, Zn, Cu
Hématite

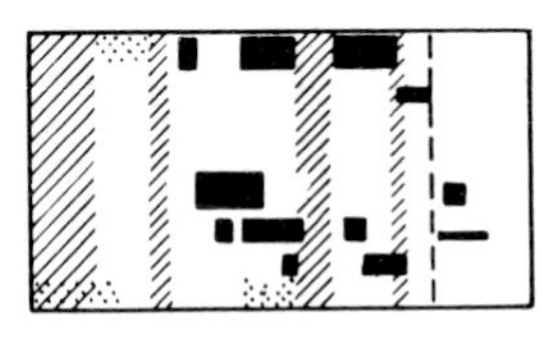

SAINT-PRIEST (B.N. 2).

Quartz
Carbonates
Fluorine
Barytine
Pechblende
Sulfure Fe
Sulf. Pb, Zn, Cu
Hématite

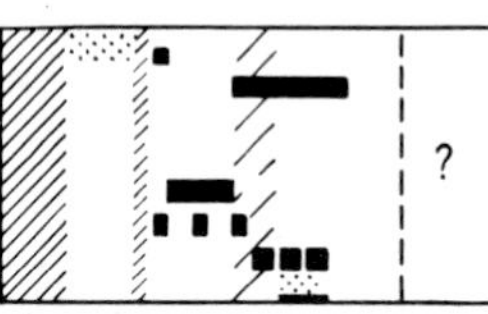

BOIS DES FAYES.

Quartz
Carbonates
Fluorine
Mispickel
Pechblende
Sulfure Fe
Sulf. Pb, Cu, Zn
Hématite

Morvan. — LA FAYE (Atelier du fond seulement).

Quartz
Carbonates
Fluorine
Barytine
Pechblende
Sulfure Fe
Sulf. Pb, Zn, Cu
Hématite

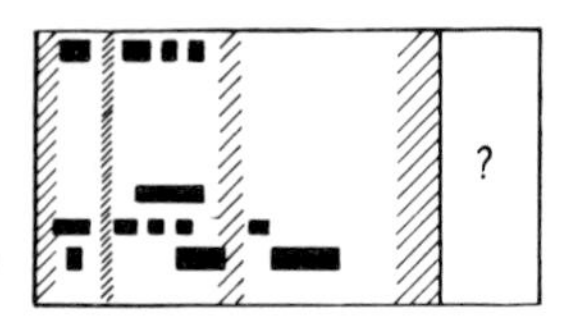

BAUZOT (La Borne-Pilot).

Quartz
Carbonates
Fluorine
Barytine
Pechblende
Sulfure Fe
Sulf. Pb, Zn, Cu
Hématite

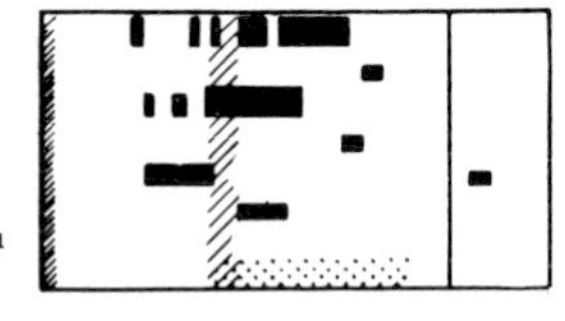

LES BROSSES.

Quartz
Carbonates
Fluorine
Barytine
Pechblende
Sulfure Fe
Sulf. Pb, Zn, Cu
Hématite

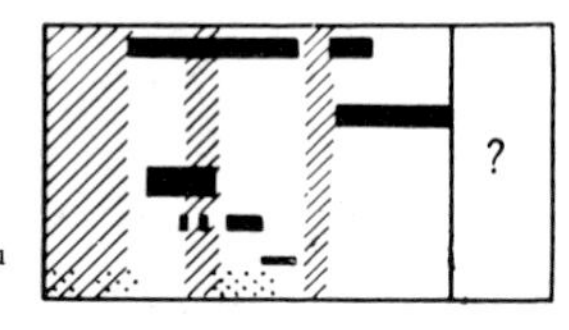

LES RUAUX.

Quartz
Carbonates
Fluorine
Barytine
Pechblende
Sulfure Fe
Sulf. Pb, Zn, Cu
Hématite

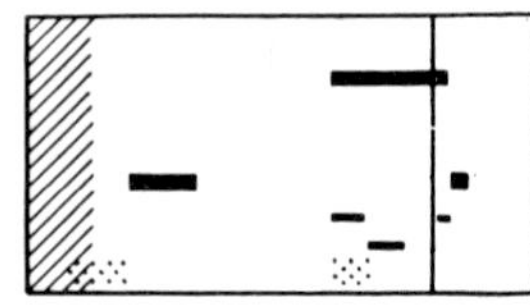

Vosges. — KRUTH.

Quartz
Arsén. Fe Ni Co
Fluorine
Barytine
Pechblende
Sulfure Fe
Sulf. Pb, Zn, Cu
Hématite

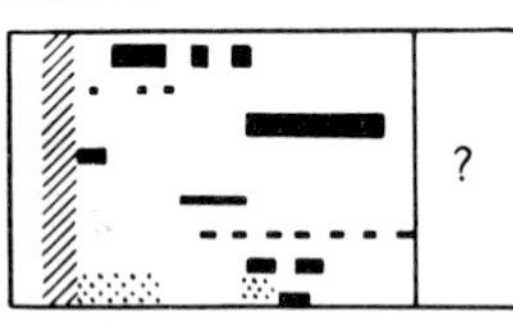

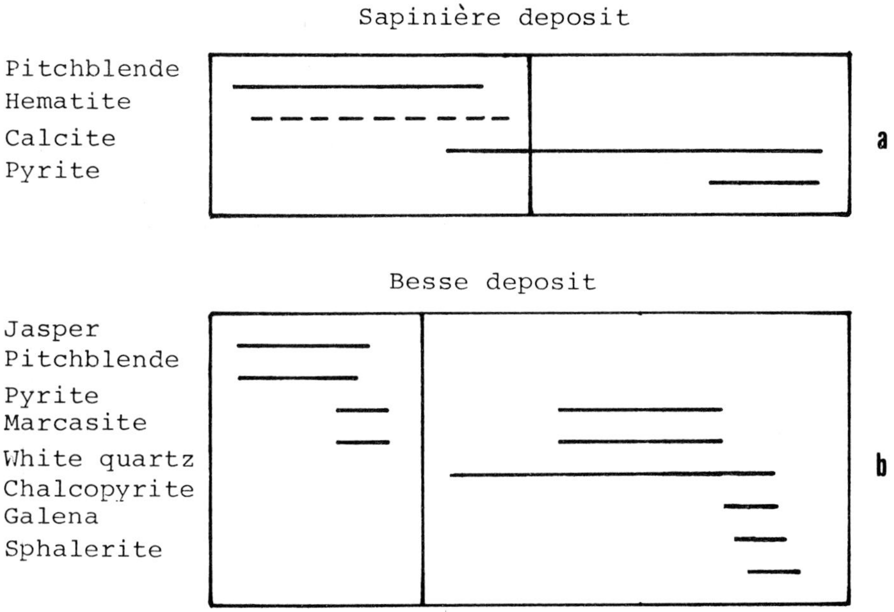

Sapinière deposit

Pitchblende
Hematite
Calcite
Pyrite

a

Besse deposit

Jasper
Pitchblende
Pyrite
Marcasite
White quartz
Chalcopyrite
Galena
Sphalerite

b

Fig. 48. Paragenesis of some hydrothermal uranium deposits in the Limousin region (modified from Geffroy and Sarcia, 1955, and Cariou, 1964; Centre d'Études Nucléaires de Saclay). (See also Table 13.)

3. Fluorite deposition.
4. Occasional late deposition of small amounts of smoky quartz + sphalerite + pyrite + galena + chalcopyrite.

In episyenite zones pitchblende is associated with minor amounts of sulfides in a carbonate gangue. Hematite is present, but is pre-pitchblende.

Table 14. Mineralogy of the Limousin region

arsenopyrite	marcasite
autunite	melnikovite
barite	parsonsite
bassetite	pitchblende
billietite	psilomelane
calcite	pyrite
chalcedony	quartz
chalcopyrite	realgar
coffinite	sabugalite
fluorite	sharpite
galena	torbernite
gummite	uranospathite
ianthinite	uranotile

Ages:

Uranium mineralization ages range from 290 to 386 m.y. (Carboniferous and Devonian) according to Ranchin (1968).

Ore Controls:

1. Hematized northwest trending fractures which post-date major regional faulting are favorable sites for uranium mineralization.
2. The richest ores are localized at lamprophyre (minette) contacts or in aplites.
3. Episyenite bodies and breccias between minette and aplite dikes are also favorable sites for uranium mineralization.

b) Forez Region

Location:

Eastern Massif Central, France (Figure 46).

Geology:

See Figure 49.

Mineralogy and Paragenesis:

See Table 13 and Figures 50a-d.

c) Morvan Region

Location:

Northeast Massif Central, France (Figure 46).

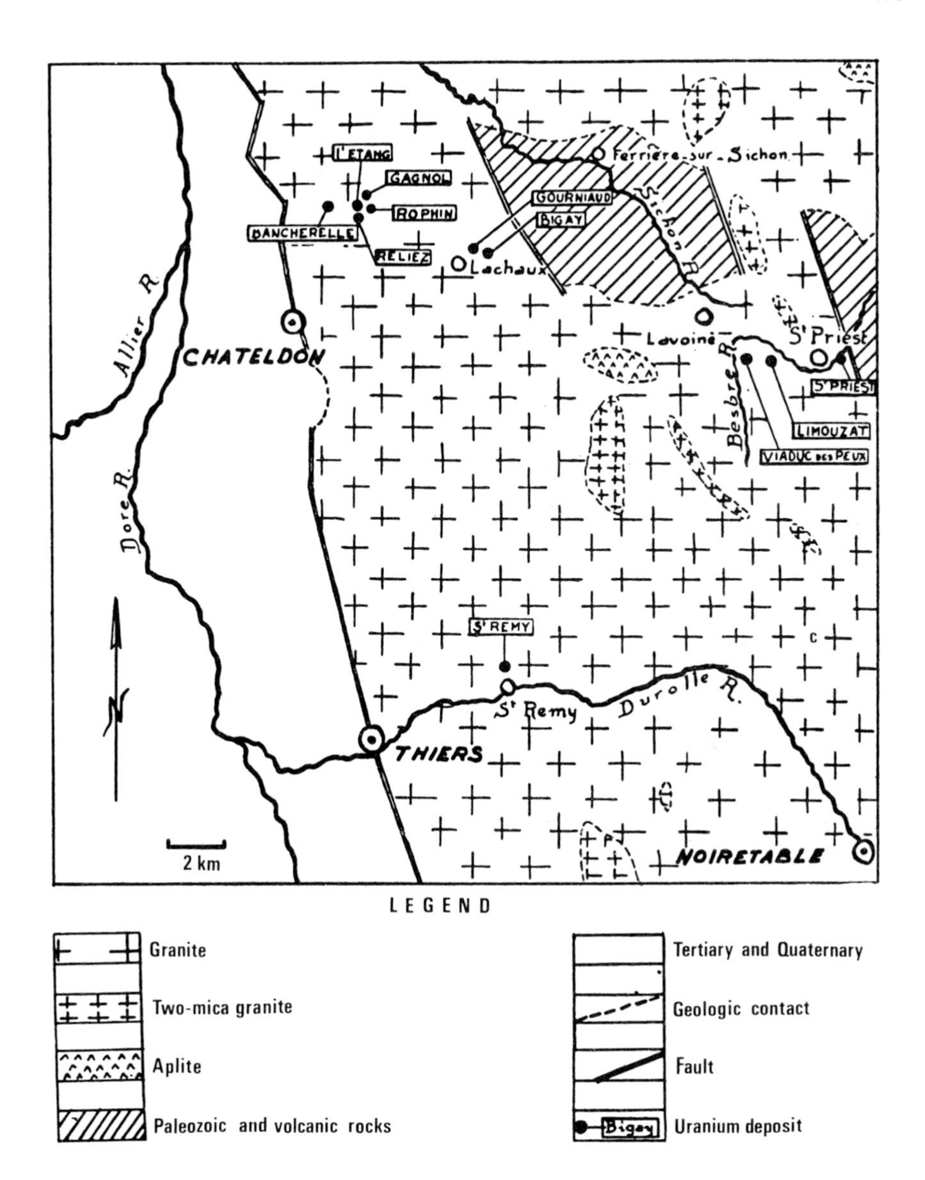

Fig. 49. Geology of the Forez region (modified from Geffroy and Sarcia, 1955; Centre d'Études Nucléaires de Saclay).

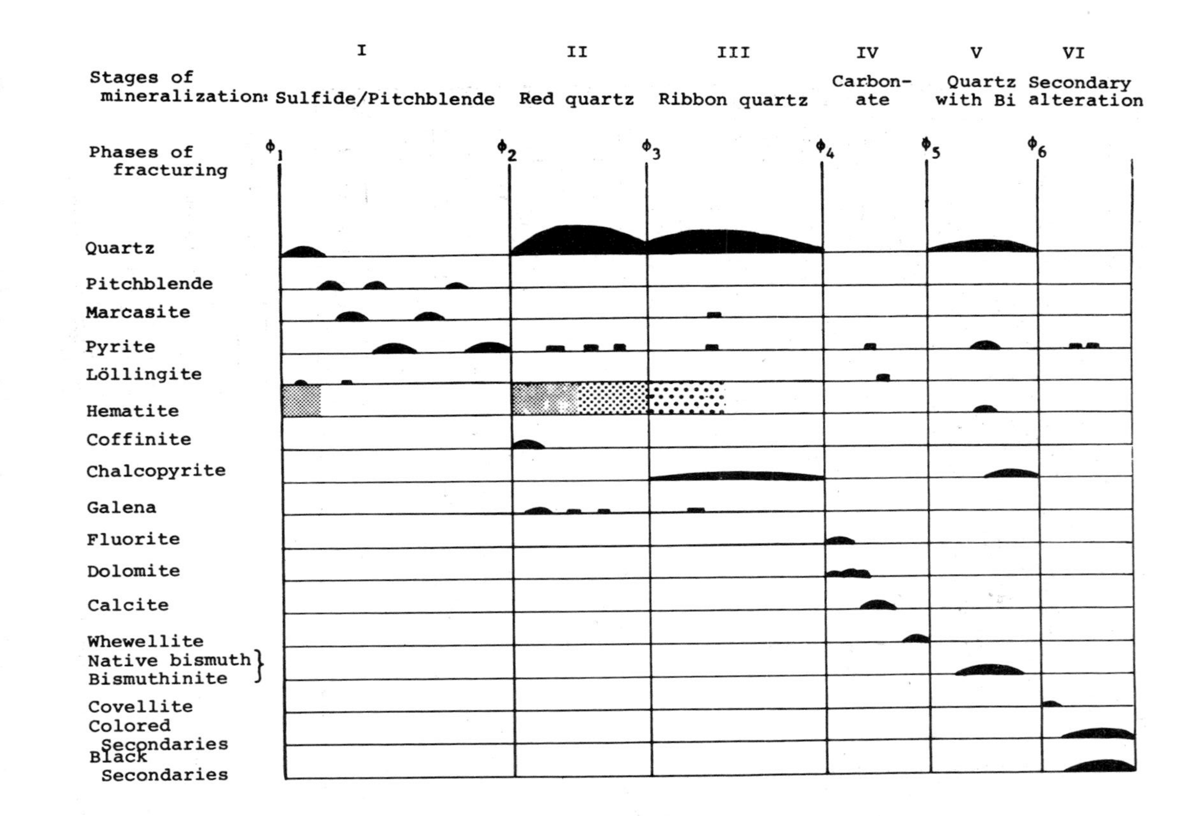

Fig. 50a. Paragenesis of the Bois Noirs (Limouzat) uranium deposit, Forez region, France (modified from Cuney, 1974).

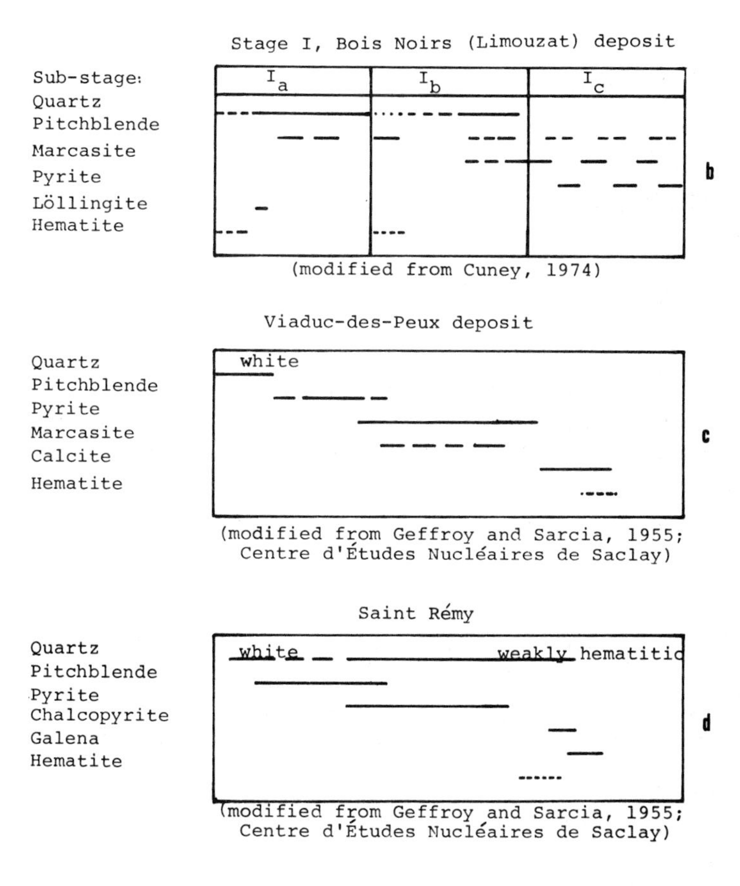

Fig. 50b—d. Paragenesis of some hydrothermal uranium deposits in the Forez region. (See also Table 13.)

Geology:

See Figure 51.

Mineralogy and Paragenesis:

See Figures 52a-g.

d) Vendée Region

Location:

Northwest France (Figure 46).

Geology:

See Figure 53.

The deposits of Vendée are unusual among French hydrothermal uranium deposits, because they occur on the margins of granites rather than within them. Mineralization is associated with mylonite and breccia zones.

Mineralogy and Paragenesis:

See Table 13.

e) Other French Uranium Deposits

The location of the Bretagne and Vosges uraniferous regions are shown in Figure 46. Paragenetic diagrams are presented in Table 13 and in Figures 54a-c for several hydrothermal uranium deposits from these regions and for two miscellaneous deposits from the Massif Central.

2) Portugal

a) Urgeirica Deposit*

Location:

Viseu-Guarda (Beira) region, northern Portugal, 7° 52′ W, 40° 32′ N (Figures 39 and 55).

Geology:

See Figure 55.

Country rocks include Cambrian schists, Hercynian granites, and basic dikes.

Mineralogy and Paragenesis:

1. Uranium occurs as microbotryoidal, sooty pitchblende.
2. Quartz-red jasper gangue.
3. Other minerals present are pyrite, galena, sphalerite, chalcopyrite, arsenopyrite, fluorite, calcite and marcasite.

*Description based on Matos Dias and Soares de Andrade (1970), except as otherwise noted.

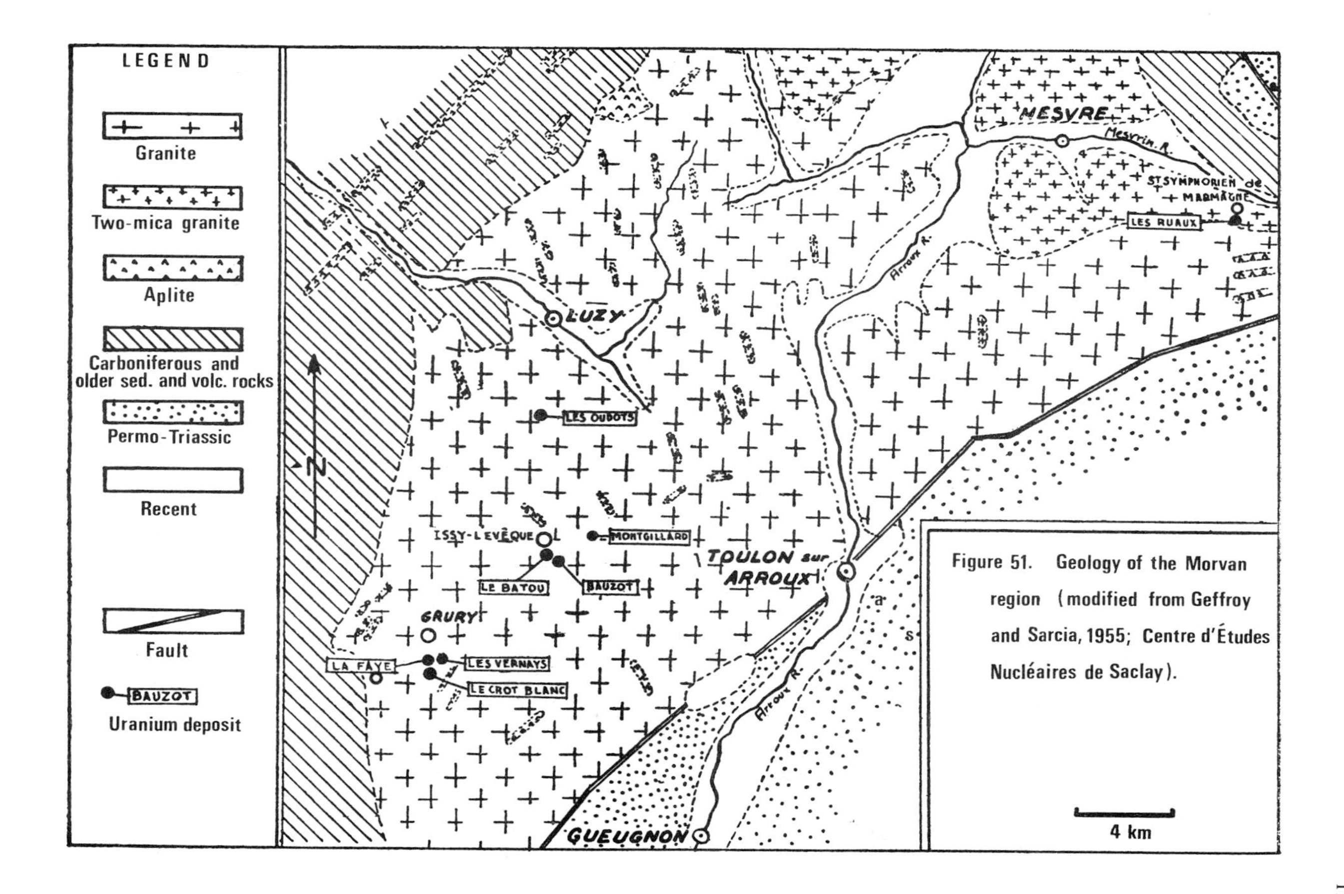

Figure 51. Geology of the Morvan region (modified from Geffroy and Sarcia, 1955; Centre d'Études Nucléaires de Saclay).

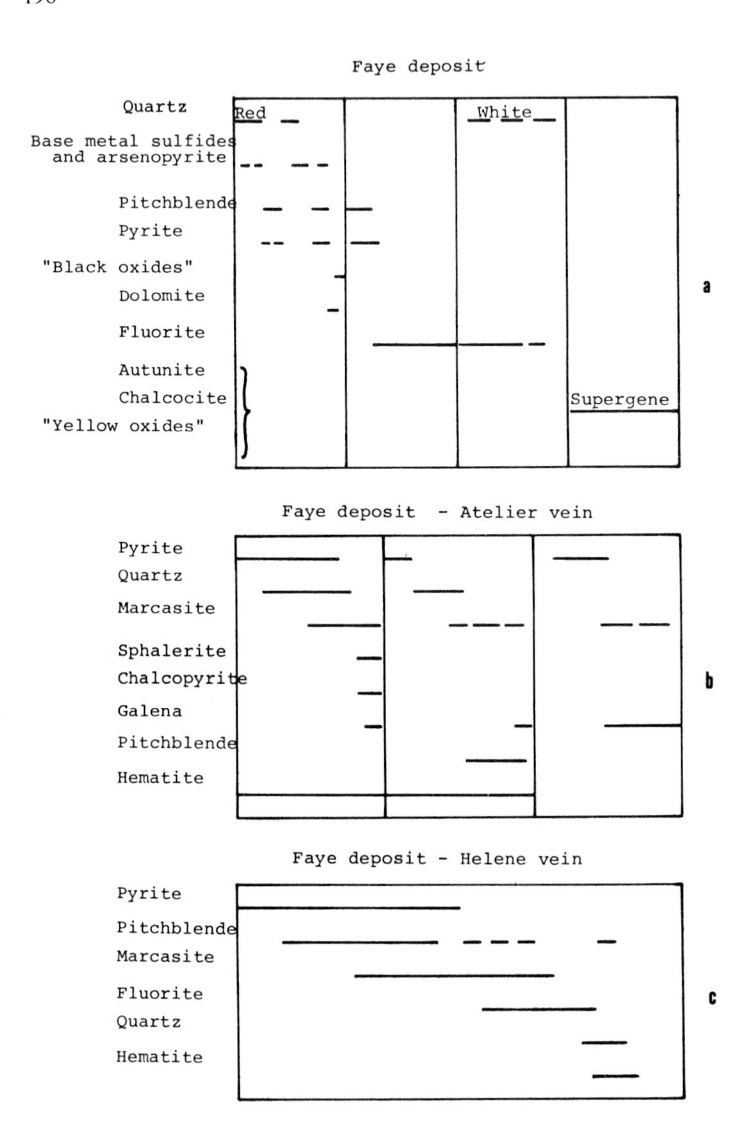

Fig. 52. Paragenesis of some hydrothermal uranium deposits in the Morvan region, France (modified from Carrat, 1962, and Geffroy and Sarcia, 1955; Centre d' Études Nucléaires de Saclay). (See also Table 13.)

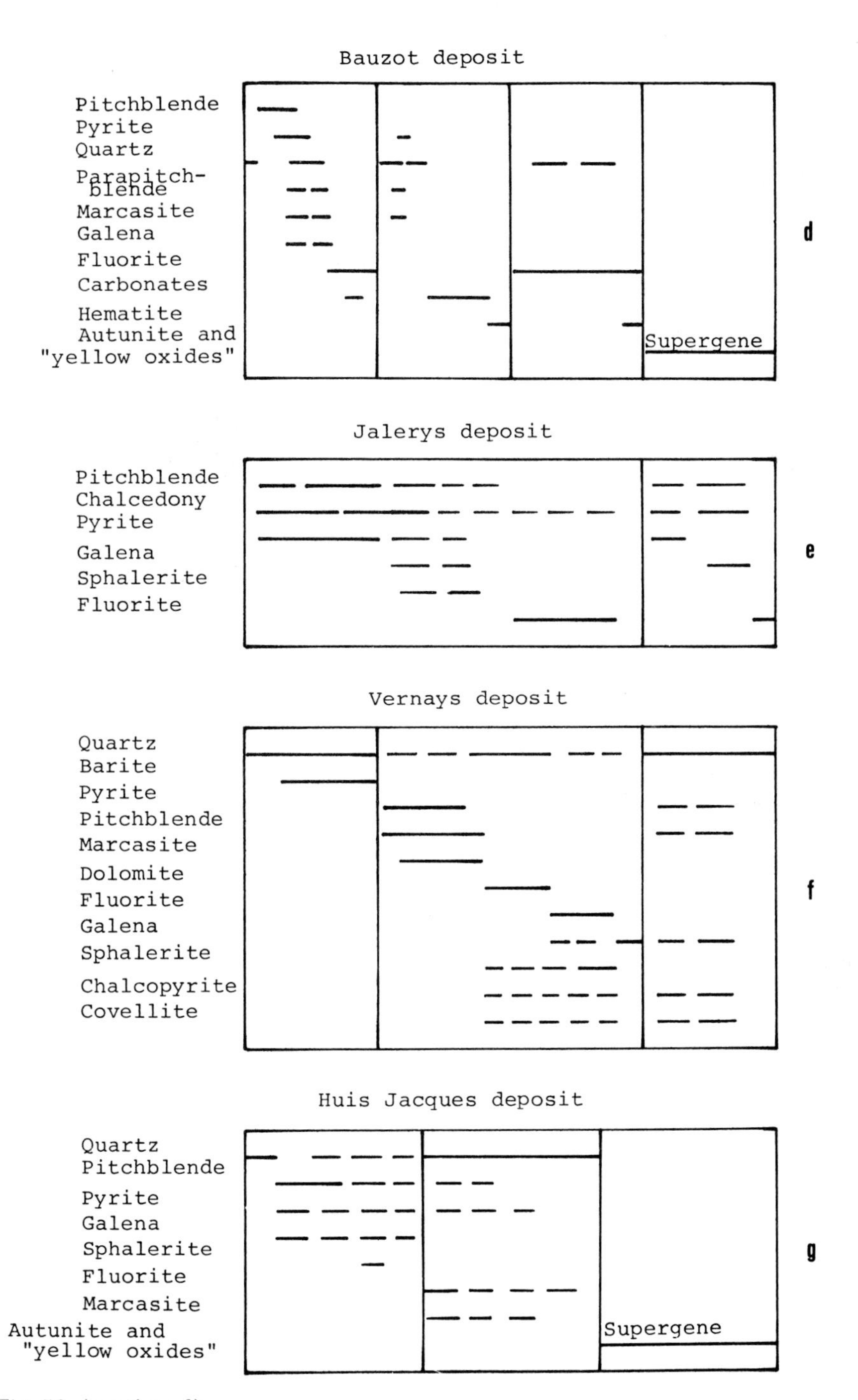

Fig. 52. (continued).

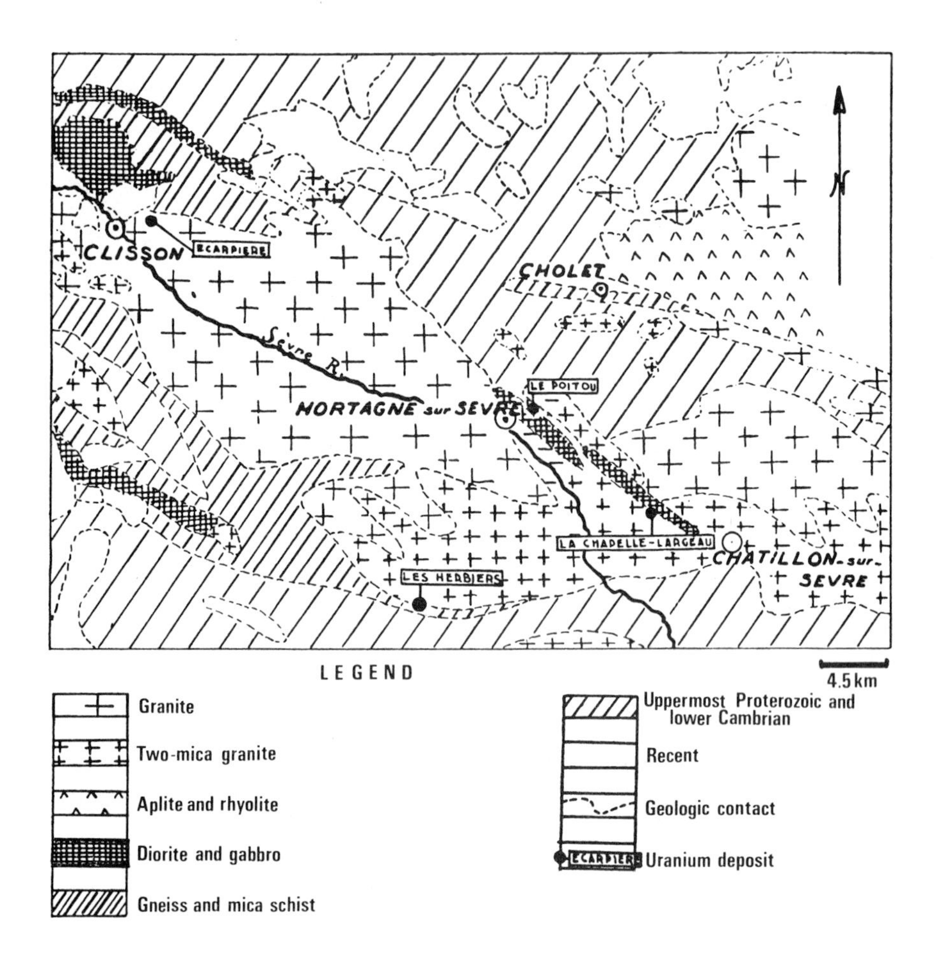

Fig. 53. Geology of the Vendée region (modified from Geffroy and Sarcia, 1955; Centre d'Études Nucléaires de Saclay).

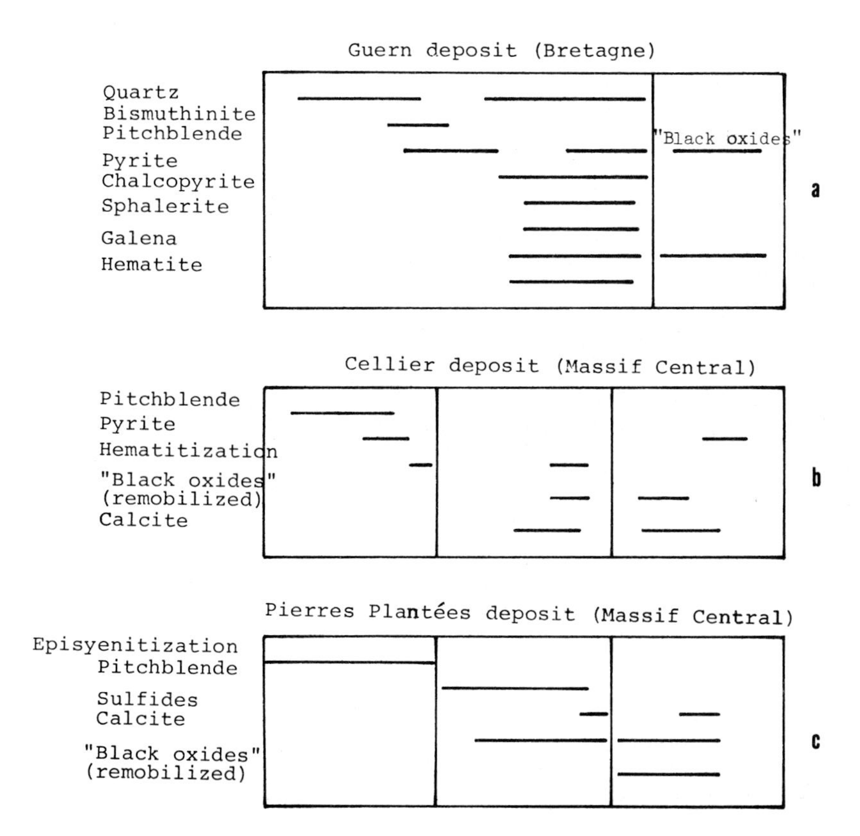

Fig. 54. Paragenesis of some miscellaneous hydrothermal deposits in France (modified from Cariou, 1964, and Germain et al., 1964; Centre d'Études Nucléaires de Saclay). (see also Table 13.)

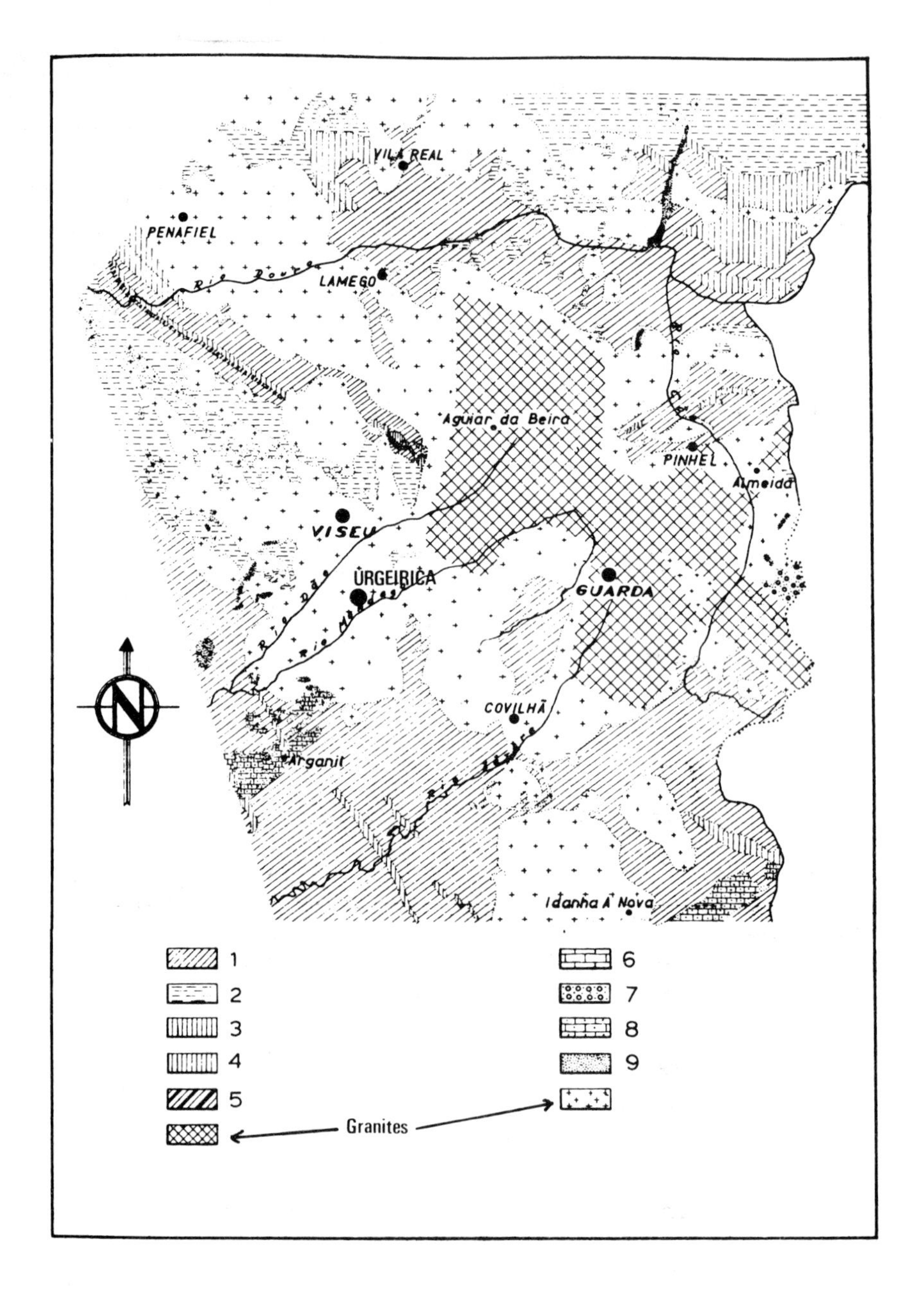

Fig. 55. Geology of the Guarda—Viseu (Beira) region, Portugal (from Lobato, 1958; United Nations). 1—2 = Paleozoic and Pre-cambrian. 1 = schist—greywacke complex; 2 = schistose crystalline rock complex; 3 = Lower Silurian; 4 = Upper Silurian; 5 = Stephanian; 6 = Upper Cretaceous; 7 = Oligocene; 8 = Miocene; 9 = Pleistocene.

Generalized sequence of vein formation:

1. White quartz with sericitic wall rock alteration.
2. Red cryptocrystalline quartz.
3. Pyrite + fluorite + sphalerite + galena.
4. Fracturing.
5. Microcrystalline quartz + pitchblende.
6. Pyrite + chalcopyrite.
7. Quartz + pyrite + pitchblende.
8. Carbonatization and kaolinization of fault breccias.

Wall Rock Alteration:

Alteration effects include sericitization, chloritization, tourmalinization, silicification, hematization and kaolinization.

Ages:

The granitic host rocks are Hercynian, and radiometric ages for vein pitchblende are Upper Cretaceous, ranging from 80 to 100 m.y.

Ore Controls:

Ore is deposited in reactivated, northeast trending Hercynian structures. These fault and shear zones are characterized by much brecciation and mylonitization.

b) Pinhel (Senhora das Fontes) Deposit*

Location:

Beira region, northeast of Guarda, northern Portugal, 40° 48′ N, 7° 4′ E (Figures 39 and 55).

Geology:

See Figure 55.

The host rocks are argillaceous and graphitic schists that occur in close proximity to granitic gneisses and granite intrusions.

Mineralogy:

The uranium ore mineral is autunite (probably formed from weathered pitchblende). The original presence of pyrite is indicated by limonite. Quartz veins contain tourmaline and some cassiterite.

Wall Rock Alteration:

Some kaolinization is reported, but this may be a supergene effect.

*Description based on Lobato and Ferrao (1958).

Ages:

The host schists are pre-Ordovician. Uranium mineralization is thought to have been deposited during the Alpine reactivation of Hercynian, northeast trending structures.

Ore Control:

The uranium ore is contained in a band of graphitic schist that occurs within granite gneiss.

3) Cornwall District*

Location:

Cornwall peninsula, southwest England, 50° 15′ N, 4° 45′ W, (Figure 39).

Geology:

Host rocks: a sequence of greenstones and metamorphosed grits and conglomerates, which has been intruded byHercynian granite.

Mineralogy, Paragenesis and Zoning:

See Table 15.

Table 15. Mineralogy of the Cornwall district

argentite	molybdenite
arsenopyrite	native bismuth
autunite	native copper
barite	native silver
bassetite	niccolite
bismuthinite	olivenite
bornite	pitchblende
cassiterite	pyrargyrite
chalcocite	pyrite
chalcopyrite	quartz
chloanthite	rammelsbergite
chlorite	rhodochrosite
cobaltite	scorodite
coffinite	siderite
cuprite	smaltite
erythrite	sphalerite
fluorite	tennantite
galena	torbernite
gold	tourmaline
hematite	uraconite
johannite(?)	vivianite
magnetite	wolframite
malachite	zeunerite
millerite	zippeite

*Description based on Darnley et al. (1965), Davidson (1956), and Park and MacDiarmid (1964).

Cornwall is one of the classical zoned ore districts. There are two major periods of mineralization: 1) Tin-copper mineralization occurring in east-west trending veins that parallel the granite contact, and 2) cobalt-nickel-uranium mineralization occurring in north-south trending veins which cut the east-west veins.

Generalized paragenetic and zoning sequence (early stages correspond to inner zones):

1. Specular hematite and cassiterite in a gangue of quartz, feldspar, mica, and tourmaline.
2. Cassiterite, arsenopyrite, and wolframite in a gangue of quartz, feldspar, mica, tourmaline, and chlorite.
3. Cassiterite, arsenopyrite, wolframite, minor stannite and molybdenite, and chalcopyrite in a gangue of quartz, tourmaline, chlorite, and fluorite.
4. Pyrite, arsenopyrite, wolframite, chalcopyrite, and minor sphalerite in a gangue of quartz, chlorite, and fluorite.
5. Minor native bismuth and bismuthinite, cobaltite, smaltite, niccolite, sphalerite, pitchblende, galena, and argentite in a gangue of quartz, chlorite, fluorite, chalcedony, and dolomite.
6. Pyrite, marcasite, siderite, bournonite, tetrahedrite, pyrargyrite, jamesonite, stibnite, hematite, and goethite in a gangue of quartz, chalcedony, barite, and calcite.
7. Minor pyrite in quartz and chalcedony.

Wall Rock Alteration:

Kaolinization and chloritization.

Ages:

Available age data suggest the existence of three periods of uranium mineralization, about 290, 225, and 50 m.y. ago.

Ore Controls:

The uranium mineralization is found at the intersections of cross fractures. Its occurrence is sporadic; pitchblende is found in lenses, not in continuous veins. The mineralization occurs in greenstone, grits, and conglomerates, but not in slaty shales.

Prospecting Guide:

Red staining of the granite is a good indicator of the presence of uranium mineralization.

Selected European References

Arnold, M. and Cuney, M., 1974, Une succession anormale de minéraux et ses conséquences sur l'exemple de la minéralisation uranifère des Bois Noirs—Limouzat (Forez, Massif Central Français): Acad. Sci. Comptes Rendus, Ser. D, *279*, 535-538.

Barbier, J., 1970, Zonalités géochimiques et métallogéniques dans le massif de Saint-Sylvestre (Limousin-France): Mineralium Deposita, *5*, 145-156.

Barbier, J., 1974, Continental weathering as a possible origin of vein-type uranium deposits: Mineralium Deposita, *9*, 271-288.

Barbier, J. and Leymarie, P., 1972, Disposition régulière de certaines minéralisations uranifères dans le granite de Mortagne (Vendée): Bur. Recherches Géol. et Minières Bull. (Ser. 2), Section 2, no. 1, 11-18.

Barbier, J. and Ranchin, G., 1969, Géochimie de l'uranium dans le Massif de Saint Sylvestre (Limousin—Massif Central Français): Sci. Terre Mém. 15, 115-157.

Bernard, J.H. and Klomínský, J., 1975, Geochronology of the Variscan plutonism and mineralization in the Bohemian Massif: Bulletin of the Geological Survey, Prague, *50*, 71-81.

Bernard, J.H. and Legierski, J., 1975, Position of primary endogenous uraninite mineralization in the Variscan mineralization system of the Bohemian Massif: Bulletin of the Geological Survey, Prague, *50*, 321-328.

Bernard, J.H., Rösler, H.J., and Baumann, L., 1968, Hydrothermal ore deposits of the Bohemian Massif: 23rd Internat. Geol. Congress, Guide to excursion 22-AC., 51 pp.

Cameron, J., 1959, Structure and origin of some uranium-bearing veins in Portugal: Junta de Energia Nuclear Tech. Paper 22.

Cariou, L., 1964, Régions médiane et Sud du Massif central: in Roubault, M., ed., *Les Minerais Uranifères Français, 3,* pt. 1, 9-162, Presses Universitaires de France, Paris.

Carlier, A., 1965, Les schistes uranifères des Vosges: in Roubault, M., ed., *Les Minerais Uranifères Français, 3,* pt. 2, 1-95, Presses Universitaires de France, Paris.

Carrat, H.G., 1962, Morvan et Autunois: in Roubault, M., ed., *Les Minerais Uranifères Français, 2,* 1-104, Presses Universitaires de France, Paris.

Carrat, H.G., 1971, Relation entre la structure des massifs granitiques et la distribution de l'uranium dans le Morvan: Mineralium Deposita, *6,* 1-22.

Carrat, H.G., 1973, Données nouvelles sur les granites uranifères du Nord-Est du Massif Central en comparaison avec ceux du Limousin et de la Vendée: in Morin, P., ed., *Les Roches Plutoniques dans leur Rapports avec les Gîtes Minéraux,* 63-76.

Chrt, J., Bolduan, H., Bernstein, K.H. and Legierski, J., 1968, Räumliche und zeitliche Beziehungen der endogenen Mineralisation der Böhmischen Masse zu Magmatismus und Bruchtektonik: Zeitschr. Angew. Geologie, *14,* 362-376.

Coppens, R., 1973, Sur la radioactivité des granites: in Morin, P., ed., *Les Roches Plutoniques dans leurs Rapports avec les Gîtes Minéraux,* 44-61.

Cuney, M., 1974, Le gisement uranifère des Bois-Noirs-Limouzat (Massif Central-France)-Relations entre minéraux et fluides: Unpub. doctoral thesis, Nancy, 174 pp.

Darnley, A.G., English, T.H., Sprake, O., Preece, E.R. and Avery, D., 1965, Ages of uraninite and coffinite from southwest England: Mineralog. Mag., *34,* 159-176.

Davidson, C.F., 1956, The radioactive mineral resources of Great Britain: Internat. Conf. Peaceful Uses of Atomic Energy, *6,* 204-206, United Nations.

Gangloff, A., 1970, Notes sommaires sur la géologie des principaux districts uranifères étudiés par la C.E.A.: *Uranium Exploration Geology,* 77-105, Internat. Atomic Energy Agency, Vienna.

Geffroy, J., 1971, Les gîtes uranifères dans le Massif Central: in Symposium Jean Jung: Géologie, géomorphologie et structure profonde du Massif Central Français, 541-579.

Geffroy, J. and Sarcia, J.A., 1954, Contribution à l'étude des pechblendes françaises: Sci. Terre, *2,* no. 1-2, 1-157.

Geffroy, J. and Sarcia, J.A., 1955, Contribution à l'étude des pechblendes françaises: Rapport C.E.A. R380, 157 pp.

Geffroy, J. and Sarcia, J.A., 1958, Quelques remarques relatives à la géochimie des filons épithermaux à pechblende: Bull. Soc. Géol. de France, 6e Ser., *8,* 531-536.

Geffroy, J. and Sarcia, J.A., 1960, Les minerais noirs: in Roubault, M., ed., *Les Minerais Uranifères Français, 1,* 1-86, Presses Universitaires de France, Paris.

Germain, C., Kervella, M. and Le Bail, F., 1964, Bretagne: in Roubault, M., ed., *Les Minerais Uranifères Français, 3,* pt. 1, 209-275, Presses Universitaires de France, Paris.

Gerstner, A., Baras, L., Pinaud, C. and Tayeb, G., 1962, Vendée: in Roubault, M., ed., *Les Minerais Uranifères Français, 2,* 293-399, Presses Universitaires de France, Paris.

Harlass, E. and Schützel, H., 1965, Zur paragenetischen Stellung der Uranpechblende in den hydrothermalen Lagerstätten des westlichen Erzgebirges: Zeitschr. Angew. Geologie, *11,* 569-581.

James, C.C., 1945, Uranium ores in Cornish mines: Royal Geol. Soc. Cornwall Trans., *17,* pt. 5, 256-268.

Jurain, G. and Renard, J.P., 1970, Remarques générales sur les caractères géochimiques des granites encaissant les principaux districts uranifères français: Sci. Terre, *15,* 195-205.

Jurain, G. and Renard, J.P., 1970, Géochimie de l'uranium dans les minéraux phylliteux et les roches du massif granitique du Mortagne-sur-Sèvre (Vendée), France: Mineralium Deposita, *5,* 354-364.

Lenoble, A. and Gangloff, A., 1958, The present state of knowledge of thorium and uranium deposits in France and the French Union: Internat. Conf. Peaceful Uses of Atomic Energy, *2,* 569-577, United Nations.

Leroy, J. and Poty, B., 1969, Recherches préliminaires sur les fluides associés à la genèse des minéralisations en uranium du Limousin (France): Mineralium Deposita, *4,* 395-400.

Leutwein, F., 1957, Alter und paragenetische Stellung der Pechblende erzgebirgischer Lagerstätten: Geologie, *6,* 797-805.

Lobato, C.P., 1958, Tectonic synthesis of Portuguese uraniferous districts: Distribution of mineralization in the Beira region: Internat. Conf. Peaceful Uses of Atomic Energy, *2,* 632-650, United Nations.

Lobato, C.P. and Ferrao, C.N., 1958, The occurrence of uranium ores in formations of pre-Ordovician schists, Pinhel, Portugal: Internat. Conf. Peaceful Uses of Atomic Energy, *2,* 651-657, United Nations.

Matos Dias, J.M. and Soares de Andrade, A.A., 1970, Uranium deposits in Portugal: in *Uranium Exploration Geology,* 129-142, Internat. Atomic Energy Agency, Vienna.

Moreau, M. and Ranchin, G., 1973, Altérations hydrothermales et contrôles tectoniques dans les gîtes filoniens d'uranium intragranitiques du Massif Central Français: in Morin, P., ed., *Les Roches Plutoniques dans leurs Rapports avec les Gîtes Minéraux,* 77-100.

Moreau, M., Poughon, A., Puibaraud, Y. and Sanselme, H., 1966, L'uranium et les granites: Chronique Mines et Recherche Minière No. 350, 47-51.

Mrna, F. and Pavlů, D., 1967, Loziska Ag-Bi-Co-Ni-As formace v Ceském masivu: Sborník geol. ved, Prague, rada LG, *9,* 7-104.

Naumov, G.B., Acheyev, B.N. and Yermolayev, N.P., 1970, Movement of hydrothermal solutions: Internat. Geology Rev., *12,* 610-618.

Paces, T., 1969, Chemical equilibria and zoning of subsurface water from Jáchymov ore deposit, Czechoslovakia: Geochim. et Cosmochim. Acta, *33,* 591-609.

Park, C.F. and MacDiarmid, R.A., 1964, *Ore Deposits:* W.H. Freeman and Co., San Francisco, 475, pp.

Pavlů, D., 1972, Ag-As-Bi-Co-Ni association in the Jáchymov ore district, Krusné Hory Mountains (abst.): 24th Internat. Geol. Congress, Section 4, 526.

Písa, M., 1966, Minerogeneze Pb-Zn-loziska v Bohutíne u Príbrami: Sborník geol. ved, Prague, LG, *7,* 164 pp.

Pluskal, O., 1970, Uranium mineralizaiton in the Bohemian Massif: in *Uranium Exploration Geology,* 107-115, Internat. Atomic Energy Agency, Vienna.

Poughon, A., 1962, Forez: in Roubault, M., ed., *Les Minerais Uranifères Français, 2,* 105-183, Presses Universitaires de France, Paris.

Ranchin, G., 1968, Contribution à l'étude de la répartition de l'uranium à l'état de traces dans les roches granitiques saines: Sci. Terre, *13,* 159-205.

Ranchin, G., 1970, La géochimie de l'uranium et la différenciation granitique dans la province du nord-Limousin: Sci. Terre Mém. 17, 483 pp.

Rösler, H.J. and Baumann, L., 1970, On the different origin of Variscan and post-Variscan ("Saxonic") mineralizations in central Europe: in Pouba, Z. and Stemprok, M., eds., *Problems of Hydrothermal Ore Deposition,* 72-77, Akadémiai Kiadó, Budapest.

Roubault, M. et al., 1969, La géologie de l'uranium dans le massif granitique de Saint-Sylvestre (Limousin, Massif Central Français): Sci. Terre Mém. 15, 213 pp.

Rumbold, R. , 1954, Radioactive minerals in Cornwall and Devon: Mining Mag., *91,* 16-27.

Ruzicka, V., 1971, Geological comparisons between East European and Canadian uranium deposits: Canada Geol. Survey Paper 70-48, 196 pp.

Sarcia, J.A., 1958, The uraniferous province of northern Limousin and its three principal deposits: Internat. Conf. Peaceful Uses of Atomic Energy, *2,* 578-591, United Nations.

Sarcia, J. and Sarcia, J.A., 1962, Gîtes et gisements du Nord Limousin: in Roubault, M., ed., *Les Minerais Uranifères Français, 2,* 185-292, Presses Universitaires de France, Paris.

Sarcia, J.A., Carrat, H.G., Poughon, A. and Sanselme, H., 1958, Geology of uranium vein deposits of France: Internat. Conf. Peaceful Uses of Atomic Energy, *2,* 592-611, United Nations.

Schneiderhöhn, H., 1941, Lehrbuch des Erzlagerstättenkunde: Verlag von Gustav Fischer, Jena, 858 pp.

Stein,P., 1952, A survey of uraniferous deposits in Cornwall: Mining Jour., *238,* no. 6079, 196-198.

Tischendorf, G., Wasternack, J., Bolduan, H. and Bein, E., 1965, Zur Lage der Granitoberfläche im Erzgebirge und Vogtland: Zeitschr. Angew. Geologie, *11,* 410-423.

Viebig, W., 1905, Die Silber-Wismutgänge von Johanngeorgenstadt im Erzgebirge: Zeitschr. Prakt. Geologie, *13,* 89-115.

IV) AFRICA

The location of some of the important hydrothermal uranium deposits in Africa are shown in Figure 56.

A) GABON*

Three important uranium deposits (Mounana, Boyindzi, and Oklo) are located on the edge of the Franceville Basin in southeast Gabon. (Figure 56). These deposits occur in Middle Precambrian clastic sediments of the Franceville Series in close proximity to Lower Precambrian granitic basement (Figure 57). The Mounana and Boyindzi deposits are clearly discordant with respect to host rock bedding, but Oklo is stratiform. Bourrel and Pfiffelmann (1972) suggest that the uranium in these deposits may have been derived from either nearby granitic basement (averaging 4 ppm uranium) or from overlying acidic tuffs (averaging 5.7 ppm uranium). Unmineralized sedimentary host rocks contain only about 2 ppm uranium.

1) Mounana Deposit

Location:

Southeast Gabon, Franceville basin, 1° 40′ S, 13° 32′ E (Figure 56).

Geology:

See Figure 57.

The Mounana deposit is located in continental sediments of the Middle Precambrian Franceville Series adjacent to granitic basement. Host rocks are conglomerates and coarse sandstones, especially sandstones cemented by asphaltic organic matter. Schists and argillites of the Franceville Series do not appear to be mineralized.

Mineralogy:

In the oxidized zone francevillite, vanuranilite and other hydrated vanadates of uranium, barium, and aluminum are found. Below the oxidized zone pitchblende with accessory coffinite occurs in association with pyrite, marcasite, and melnicovite. Some galena, sphalerite, chalcopyrite and chalcocite are also found. Dominant gangue minerals are calcite and barite. In the unoxidized zone vanadium minerals occur in different mineralized structures than the uranium minerals. Vanadium minerals from this zone include karelianite, montroseite, roscoelite, duttonite, and corvusite.

*The description of the Gabon deposits is based on Bourrel and Pfiffelmann (1972), except as otherwise noted.

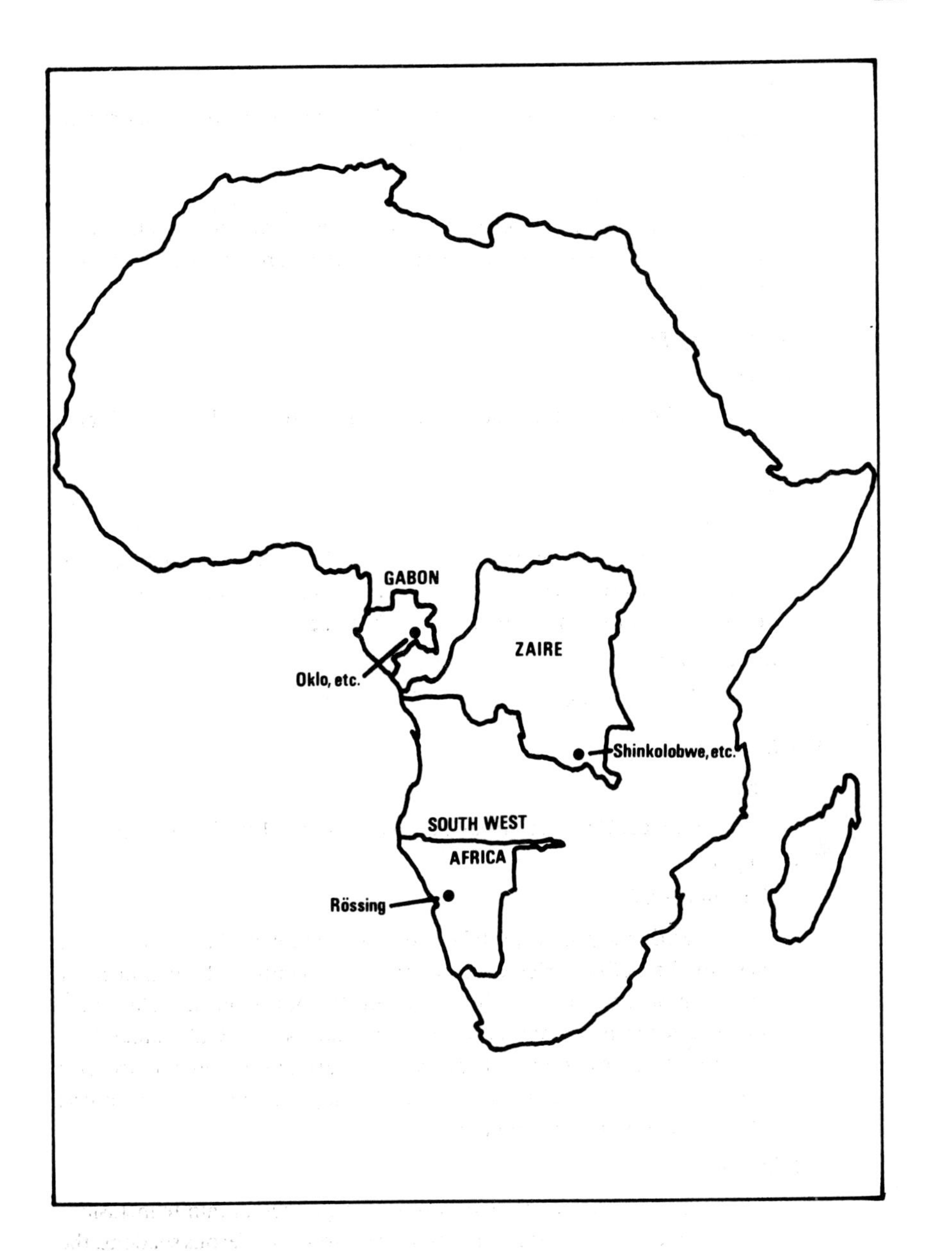

Fig. 56. Locations of some important uranium deposits in Africa.

Age:

Franceville Series rocks from the Mounana area are approximately 1740 m.y. old (Gangloff, 1970).

Ore Controls:

Both structure (brecciation associated with faulting) and host rock lithology (sediments rich in organic matter) are thought to be ore controls.

2) Boyindzi Deposit

Location:

Southeast Gabon, Franceville Basin, 1° 40′ S, 13° 32′ E (Figure 56).

Geology:

See Figure 57.

The Boyindzi deposit occurs in Middle Precambrian conglomerates and coarse sandstones (commonly cemented by organic matter) of the Franceville Series adjacent to granitic basement.

Ore Controls:

Primarily faults.

3) Oklo Deposit

Location:

Southeast Gabon, Franceville basin, 1° 41′ S, 13° 32′ E (Figure 56).

Geology:

See Figure 57.

The Oklo deposit is stratabound. The mineralized bed is close to the base of the pelitic series which overlies the sandy and conglomeratic beds that host the nearby Mounana and Boyindzi deposits. The pelitic series is richer in organic matter than the sandy series. Oklo mineralization locally is less than 100 m from the granitic basement. Isotopic evidence indicates that the Oklo uranium deposit operated as a natural atomic reactor approximately 1.7 b.y. ago.

Mineralogy:

In the ore zone uranium occurs as fine-grained uraninite inclusions in carbonaceous cementing material. In poorer uraniferous sections, the proportion of phyllitic cement increases and coffinite appears. Associated minerals are pyrite; chalcopyrite, galena, and sphalerite. The oxidized zone contains many U^{+6} minerals, but francevillite and wölsendorfite are most important.

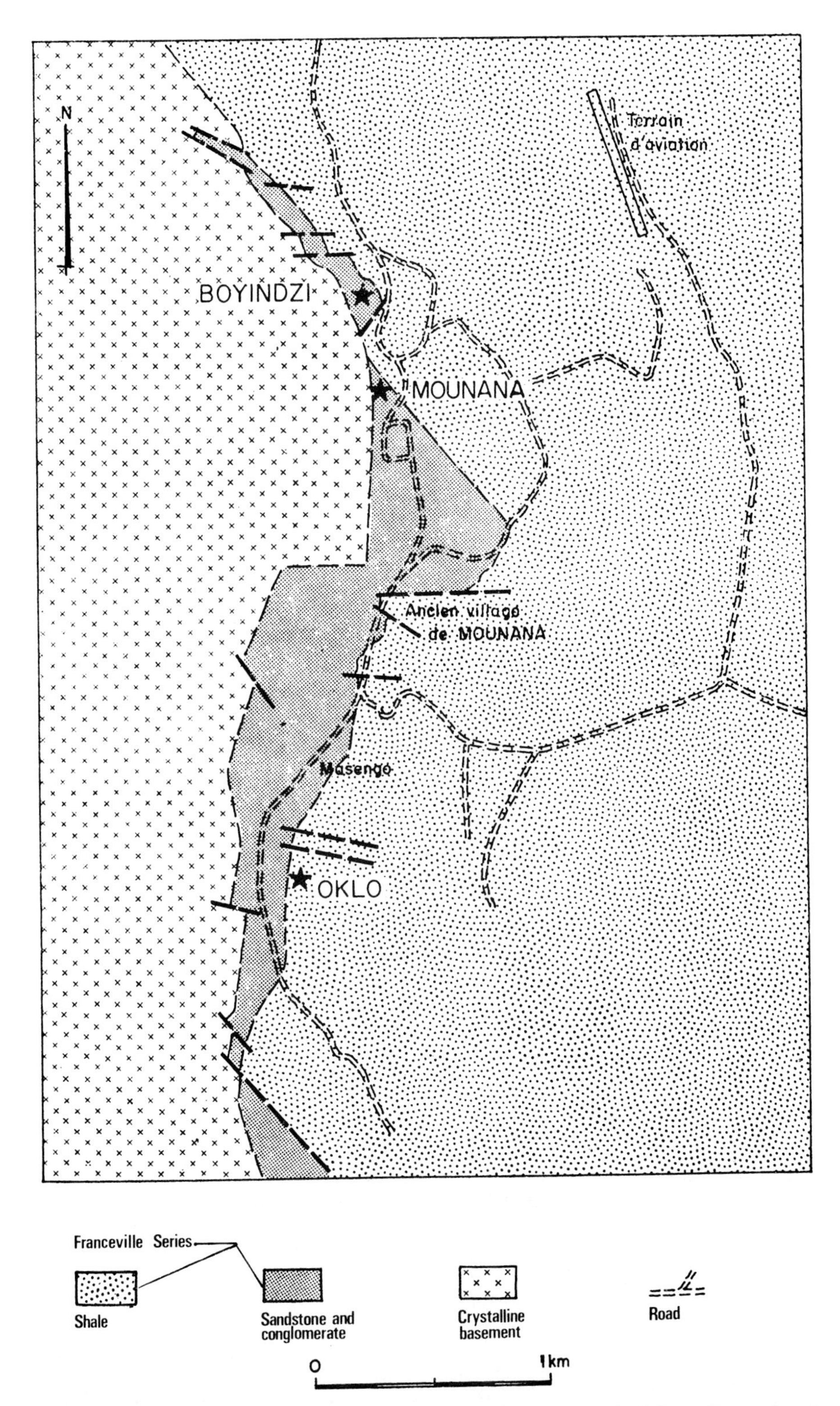

Fig. 57. Geology of the uranium deposits of Gabon (modified from Bourrel and Pfiffelmann, 1972).

Ore Controls:

Structural control is indicated by the richly mineralized areas associated with fold axes. There also appears to be a relation between mineralization and both sedimentary lithology and bed thickness.

B) SOUTH WEST AFRICA

Rössing Mine *

Location:

South West Africa, Namib desert, 55 km northeast of Swakopmund, 22° 25′ S, 15° 01′ E. (Figure 56).

Introductory Note:

Because of the current interest in the Rössing mine as a type example of a "porphyry uranium deposit", the following description has been included in this book even though published data probably preclude a strictly hydrothermal origin for the uranium mineralization.

Geology:

See Figure 58.

The Rössing mine area is underlain by highly folded, metamorphosed, and partially granitized Precambrian rocks of varied (but mostly shallow marine) sedimentary lithologies. Marbles, calc-silicate rocks, quartzites, and various quartzo-feldspathic, biotitic metasediments are most common. These rocks are part of the Damara Supergroup. In the Rössing area, the sequence composed of the upper portion of the Etusis Formation and the overlying Khan and Rössing Formations was particularly favorable for the intrusion of the alaskitic pegmatites which contain the uranium mineralization. Pegmatite intrusion is locally so intense that the host rocks may occur only as xenoliths.

In the Rössing area pegmatites, consist largely of anhedral smoky quartz, subhedral salmon pink microcline, and accessory black biotite. Structural relations indicate that pegmatites in the Rössing area have ages ranging from early syntectonic to post-tectonic. The uraniferous pegmatites are thought to be a product of granitization processes associated with high grade regional metamorphism and tectonism.

The Rössing uranium deposit consists of a large number of narrow, uraniferous zones within alaskitic pegmatite intrusions. The ore zone is irregular in shape, and has a diameter of about 700 m. Uranium minerals occur mostly as fine-grained disseminations in pegmatites, but small stringers of secondary uranium minerals are sometimes seen. There are no known pitchblende veins in the Rössing area.

*Except as otherwise noted, this description is based on Smith (1965) and Berning et al. (1976).

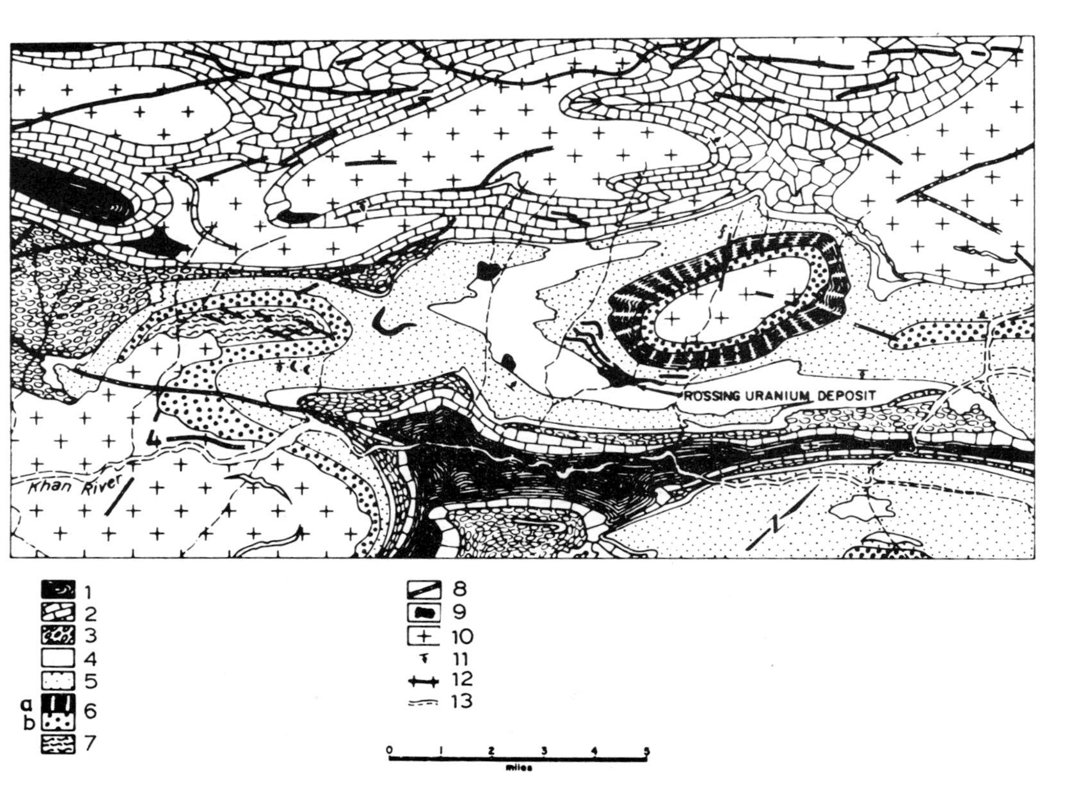

Fig. 58. Geology of the Rössing uranium deposit (from Berning et al., 1976). 1—7 = Damara Supergroup: 1 = Khomas Formation; 2 = Welwitch Formation; 3 = Chuos Formation; 4 = Rossing Formation; 5 = Khan Formation; 6 = Etusis Formation (a = biotite gneiss; b = feldspathic quartzite). 7 = Abbabis Formation; 8 = dolerite dyke (Post-karroo); 9 = pegmatitic granite (only major occurrences shown); 10 = granite—gneiss and granite (with sedimentary remnants). 9, 10 = Syn-to Post-Orogenic. 11 = strike and dip of foliation; 12 = railway; 13 = rivers.

Uranium values as high as 0.55 weight % U_3O_8 were encountered during early exploration work, but such high grade zones are small and extremely patchy. Several million tons of low grade (0.05 weight % U_3O_8) mineralization have been proven (von Backström, 1970). Ruzicka (1975) estimates that the Rössing mine contains reserves of about 150,000 tons of U_3O_8 in ore with an average grade of 0.7 lb./ton.

Mineralogy:

The major uranium mineral at Rössing is uraninite. It occurs as minute grains ($\leq$0.3mm in diameter), either as inclusions or along cracks in quartz, feldspar, and biotite or as free interstitial grains in pegmatite. Hiemstra (1969) reports that uraninite constitutes 55% of all the radioactive minerals present. Unleachable primary betafite makes up $<$5%, and secondary uranium minerals the remaining 40% of the radioactive minerals present in near-surface samples. Ruzicka (1975) states that the secondary minerals represent an enrichment of the primary uranium content. If this is true, then it is the supergene enrichment that makes the Rössing uranium occurrence an ore deposit. Secondary uranium minerals include rössingite, beta-uranophane, metatorbernite, metahaiweeite, uranophane, carnotite, thorogummite and gummite. Much of the secondary mineralization is present along joints, cracks, cleavages, and grain boundaries in quartz and feldspar, and between flakes of biotites. Most of the uranium minerals present are easily leachable. Other minerals associated with the uranium mineralization are zircon, apatite, sphene, and monazite (often closely associated with uraninite), minor pyrite, chalcopyrite, bornite, molybdenite, arsenopyrite, ilmenite (intergrown with magnetite), and rarely fluorite, and hematite (as inclusions in biotite and feldspar).

Age:

Nicolaysen (1962) reported radiometric ages for uraninite, davidite, monazite, biotite, yttrotantalite, and yttrocolumbite from seven pegmatites and one marble of the Rössing area. These data indicate an age of 510 ± 40 m.y. for the Salem granite, the uraniferous pegmatites, and the several events of regional metamorphism and deformation.

Ore Controls:

Uranium mineralization is confined to intrusive bodies of alaskitic pegmatite. Within the uraniferous zone enrichment is present in biotite-rich selvages of the alaskite. Regionally, uraniferous pegmatites of the lower Khan River gorge are conspicuously associated with bands of biotite schist in Lower Damara rocks. Similarly, where pegmatites cut biotite-rich members of the Abbadis Formation (pre-Damara basement), they are commonly uraniferous.

The dolomitic marble of the lower part of the Damara Supergroup may correlate with the copper and uranium rich Upper Roan dolomites of Zaire and Zambia (see Nicolayson, 1962). In this regard it is interesting that in the Rössing area, Lower Damara calc-silicate bands commonly contain accessory pyrite, pyrrhotite, bornite, and chalcopyrite.

C) ZAIRE*

Hydrothermal uranium mineralization is generally associated with dolomitic rocks similar to those of the Zambian Copperbelt. Figure 59 shows the location of the principal uranium occurrences in Zaire. This figure also shows the regional structure of the Lufilian arc as outlined by the distribution of outcrops of the Schisto-Dolomite, the favorable host rock for the uranium deposits. Although granitic basement crops out at many places in the Lufilian arc region, there is no clear spatial relationship between the distribution of uranium deposits and granitic rocks.

1) Shinkolobwe Deposit

Location:

Zaire, 130 km northwest of Elizabethville, 11° 13′ S, 26° 40′ E (Figures 56 and 59).

Geology:

The deposit is located in a faulted anticline. Most of the uranium ore occurs in the Schisto-Dolomite of the Precambrian Série des Mines. The uranium host rocks at Shinkolobwe are lithologically very similar to the altered carbonate rocks of the Zaire (Katanga) copper deposits. Uranium mineralization occurs in short, erratic veins.

Mineralogy and Paragenesis:

See Table 16.

At Shinkolobwe copper is present in only minor amounts, whereas cobalt and nickel are abundant. In most other Copperbelt deposits copper clearly predominates over cobalt and nickel.

Generalized sequence of deposition:

1. Magnesite.
2. Uraninite.
3. Pyrite, molybdenite and monazite.
4. Cobalt-nickel sulfides (vaesite, cattierite and siegenite).
5. Chalcopyrite.

*The descriptions of the Zaire deposits are based on Cahen et al. (1971), Derriks and Oosterbosch (1958), and Derriks and Vaes (1956).

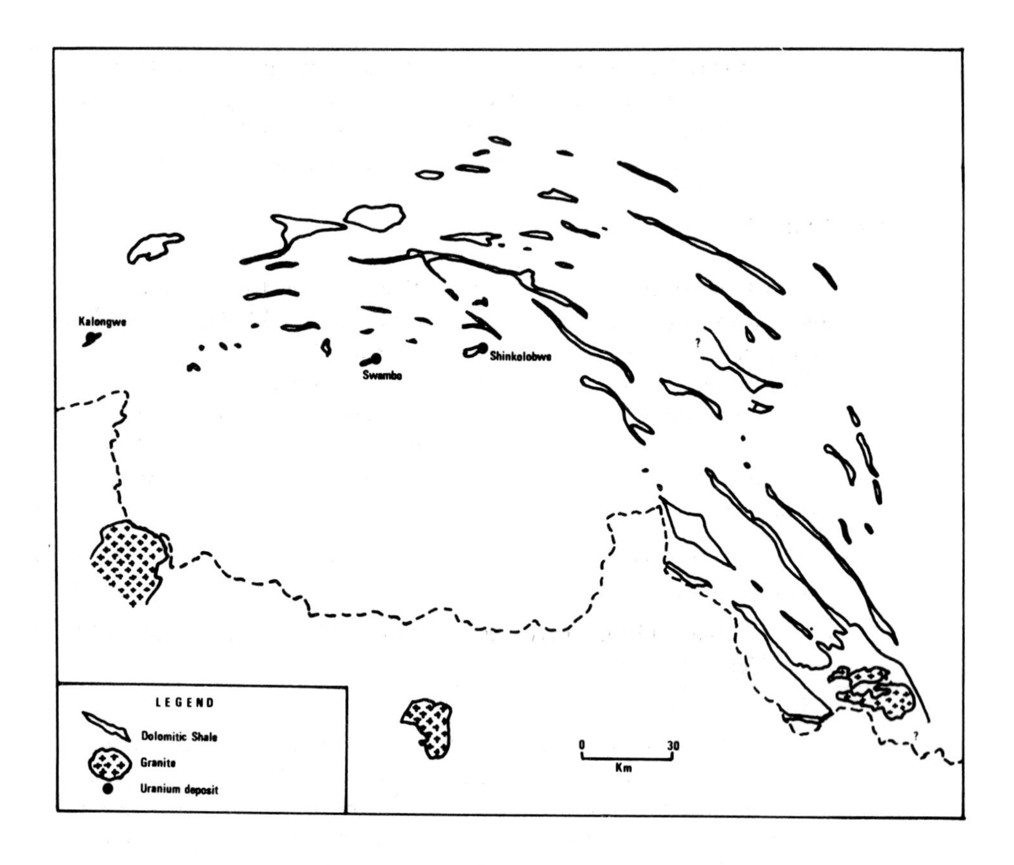

Fig. 59. Location of major hydrothermal uranium deposits in the Dolomitic Shale of the Lufilian arc, Zaire (redrawn from Derriks and Oosterbosch, 1958, United Nations). 1 = dolomitic shale; 2 = granite; 3 = uranium deposit.

Table 16. Mineralogy of the Shinkolobwe deposit (from Derriks and Vaes, 1956).

aragonite	metatorbernite
becquerelite	millerite
billietite	molybdenite
bornite	monazite
calcite	native copper
cattierite	parsonite
chalcopyrite	pyrite
chlorite	quartz
covellite	renardite
cuprosklodowskite	richetite
curite	rutherfordine
dewindtite	saleite
diderichite	schoepite
digenite	sharpite
dolomite	siegenite
dumontite	sklodowskite
fourmarierite	soddyite
galena	studtite
garnierite	torbernite
gold	umangite
heterogenite	uraninite
ianthinite	uranophane
kasolite	vaesite
linnaeite	vandenbrandeite (or uranolepidite)
magnesite	vandendriesscheite
masuyite	wulfenite
melonite	

Notes:
1) Vaesite is a characteristic mineral of the uraninite zone.
2) Selenides are commonly present in the ore.
3) Gold is found in uraninite and sulfides from the central part of the mine.
4) The dominant gangue mineral is dolomite, especially with the nickel-cobalt sulfides.
5) Tourmaline and apatite were deposited with early mineralization; i.e. before uraninite.

Zoning:

Cobalt and nickel sulfides occur within, and extend beyond, the uraniferous zone. The nickel to cobalt ratio is high (3:1) near the uranium zone and progressively decreases to a value of 1:3 away from the uranium zone.

Wall Rock Alteration:

Chloritization of the Schisto-Dolomite host rocks.

Ages:

Uranium mineralization ages of 620 ± 20 m.y., 670 ± 20 m.y. and ≥706 m.y. are given by Cahen et al. (1971).

Ore Controls:

Mineralization is located where brecciated fault and fracture zones intersect rocks of the Mine Series (Schisto-Dolomite or Roan System); ore also occurs along fold axes. Most ore is limited to the siliceous dolomitic portion of the Mine Series, occurring especially in thin-bedded quartzites.

2) Swambo Deposit

Location:

Zaire, between Elizabethville and Kolwezi, west of Shinkolobwe, 26° 10′ E, 11° 5′ S (Figures 56 and 59).

Geology:

Like Shinkolobwe, Swambo is located in a faulted anticline. The mineralization occurs in rocks of the Precambrian Mine Series which are found in the core of the anticline or as scales (repeated sections) in a nappe.

Mineralogy and Paragenesis:

Minerals present include:

bastnaesite (or xenotime)
chalcopyrite
chlorite
chrysocolla
covellite
curite
digenite
dolomite
gold
heterogenite
kasolite
libethenite
limonite
malachite
monazite
metatorbernite
pyrite
quartz
siegenite
sklodowskite
soddyite
vaesite
uraninite
uranophane

Generalized sequence of deposition:

1) Uraninite.
2) Pyrite and monazite, accompanied by hematization, silicification, chloritization, and recrystallization of dolomite.
3) Cobalt-nickel sulfides (vaesite and siegenite).
4) Chalcopyrite.

Zoning:

Iron, cobalt, and nickel sulfides are found both within and outside the uraniferous zone.

Wall Rock Alteration:

Local chloritization, hematization, and silicification.

Ages:

Uraninite ages of 992, 710, 697 and 682 m.y. are reported by Derriks and Oosterbosch (1958); Cahen et al. (1971) dated the mineralization at 670 m.y.

Ore Controls:

Uranium mineralization appears more to be localized by faults than by the lithology of the host rocks. Much brecciation is associated with the fault zones.

3) Kalongwe Deposit

Location:

Zaire, southwest of Kolwezi and west of Swambo and Shinkolobwe, 25° 15′ E, 11° S (Figure 59).

Geology:

The Kalongwe deposit is located in a faulted anticline.The host rocks are part of the Precambrian Schisto-Dolomite.

Mineralogy and Paragenesis:

Minerals present include:

carrollite
chalcocite
chalcopyrite
chlorite
chrysocolla
covellite
cuprosklodowskite
dolomite
heterogenite
malachite
olivenite
pyrite
quartz
torbernite
uraninite
uranophane
vandenbrandeite

Notes:
1) Pyrite and chalcocite, the dominant sulfides, occur in dolomite gangue.
2) There are no nickel sulfides, but some nickel occurs in carrol'-lite.
3) This deposit is unusually rich in copper.
4) No selenium or rare earth minerals have been reported.

Generalized sequence of deposition:

1) Uraninite.
2) Pyrite and chlorite.
3) Carrollite.
4) Bornite and chalcopyrite.

Wall Rock Alteration:

Silicification and recrystallization of dolomite.

Age:

An age of 620 m.y. is given by Cahen et al. (1971) for the uraninite of the Kalongwe deposit.

Selected African References

Backström, J.W. von, 1970, The Rössing uranium deposit near Swakopmund, South West Africa: in *Uranium Exploration Geology,* 143-150, Internat. Atomic Energy Agency, Vienna.

Berning, J., Cooke, R., Hiemstra, S.A. and Hoffman, U., 1976, The Rössing uranium deposit, South West Africa: Econ. Geology, *71,* 351-368.

Bourrel, J. and Pfiffelmann, J.P., 1972, La province uranifère du bassin de Franceville (République Gabonaise): Mineralium Deposita, *7,* 323-336.

Cahen, L., François, A. and Ledent, D., 1971, Sur l'âge des uraninites de Kambove Ouest et de Kamoto Principal et révision des connaissances relatives aux minéralisations uranifères du Katanga et du Copperbelt de Zambia: Soc. Géol. Belgique Annales, *94,* 185-198.

Derriks, J.J. and Oosterbosch, R., 1958, The Swambo and Kalongwe deposits compared to Shinkolobwe: contribution to the study of Katanga uranium: Internat. Conf. Peaceful Uses of Atomic Energy, *2,* 663-695, United Nations.

Derriks, J.J. and Vaes, J.F., 1956, The Shinkolobwe uranium deposit: current status of our geological and metallogenic knowledge: Internat. Conf. Peaceful Uses of Atomic Energy, *6,* 94-128, United Nations.

Drozd, R.J., Hohenberg, C.M. and Morgan, C.J., 1974, Heavy rare gases from Rabbit Lake (Canada) and the Oklo mine (Gabon): Natural spontaneous chain reactions in old uranium deposits: Earth and Planetary Sci. Letters, *23,* 28-33.

Gangloff, A., 1970, Notes sommaires sur la géologie des principaux districts uranifères étudiés par la C.E.A.: in *Uranium Exploration Geology,* 77-105, Internat. Atomic Energy Agency, Vienna.

Hagemann, R., Lucas, M., Nief, G. and Roth, E., 1974, Mésures isotopiques du rubidium et du strontium et essais de mésure de l'âge de la minéralisation de l'uranium du réacteur naturel fossile d'Oklo: Earth and Planetary Sci. Letters, *23,* 170-188.

Hiemstra, S.A., 1969, The determination of the relative amounts and association of uraninite and secondary uranium-bearing minerals in two composite samples from the S.J. area at Rössing: Nat. Inst. Met. Confidential Rept. No. 495.

Nicolaysen, L.O., 1962, Stratigraphic interpretation of age measurements in southern Africa, in Engel, A.E.J. and Leonard, B.F., eds., *Petrologic Studies:* A volume in honor of A.F. Buddington, 569-598, Geol. Soc. America.

Ruzicka, V., 1975, New sources of uranium? Types of uranium deposits presently unknown in Canada: Canada Geol. Survey Paper 75-26, 13-20.

Smith, D.A.M., 1965, The geology of the area around the Khan and Swakop Rivers in South West Africa: South Africa Geol. Survey, South West Africa Ser. Mem. 3, 113 pp.

Appendix I. References to data on the uranium content of granitic rocks.

Adams, J.A.S., 1954, Uranium and thorium contents of volcanic rocks: in Faul, H., ed., *Nuclear Geology,* 89-98, John Wiley and Sons, Inc., New York.

Barbier, J., 1968, Altération chimique et remaniement de l'uranium dans le granite à deux micas des Monts de Blond (Limousin, France): Sci. Terre, *13*, 359-378.

Barbier, J. and Ranchin, G., 1969, Géochimie de l'uranium dans le Massif de Saint Sylvestre (Limousin—Massif Central Français): Sci. Terre Mém. 15, 115-157.

Barbier, J. and Ranchin, G., 1969, Influence de l'altération météorique sur l'uranium à l'état de traces dans le granite à deux micas de St. Sylvestre: Geochim. et Cosmochim. Acta, *33,* 39-47.

Barbier, J., Carrat, H.G. and Ranchin, G., 1967, Présence d'uraninite en tant que minéral accessoire usuel dans les granites à deux micas uranifères du Limousin et de la Vendée: Acad. Sci. Comptes Rendus, Ser. D, *264,* 2436-2439.

Berzina, I.G., Yeliseyeva, O.P. and Popenko, D.P., 1974, Distribution relationships of uranium in intrusive rocks of northern Kazakhstan: Internat. Geology Rev., *16,* 1191-1204.

Bunker, C.M. and Bush, C.A., 1973, Radioelement and radiogenic heat distribution in drill hole UCe-1, Belmont Stock, central Nevada: U.S. Geol. Survey Jour. Research, *1*, 289-292.

Burwash, R.A. and Cumming, G.L., 1976, Uranium and thorium in the Precambrian basement of western Canada: I. Abundance and distribution: Canadian Jour. Earth Sci., *13,* 284-293.

Butler, A.P., Jr., 1956, White Mountain plutonic series, New Hampshire: U.S. Atomic Energy Comm. Rept. TEI-640, 297-299.

Davis, J.D. and Guilbert, J.M., 1973, Distribution of the radioelements K, U, and Th in selected porphyry copper deposits: Econ. Geology, *68,* 145-160.

Gottfried, D., 1956, Distribution of uranium in igneous complexes—Precambrian granites of the Front Range of Colorado: U.S. Atomic Energy Comm. Rept. TEI-640, 292-297.

Heier, K.S. and Rhodes, J.M., 1966, Thorium, uranium and potassium concentrations in granites and gneisses of the Rum Jungle complex, Northern Territory, Australia: Econ. Geology, *61,* 563-571.

Jurain, G. and Renard, J.P., 1970, Géochimie de l'uranium dans les minéraux phylliteux et les roches du massif granitique du Mortagne-sur-Sèvre (Vendée), France: Mineralium Deposita, *5*, 354-364.

Jurain, G. and Renard, J.P., 1970, Remarques générales sur les caractères géochimiques des granites encaissant les principaux districts uranifères français: Sci. Terre, *15*, 195-205.

Killeen, P.G. and Heier, K.S., 1975, A uranium and thorium enriched province of the Fennoscandian shield of southern Norway: Geochim. et Cosmochim. Acta, *39*, 1515-1524.

Kolbe, P. and Taylor, S.R., 1966, Major and trace element relationships in granodiorites and granites from Australia and South Africa: Contr. Mineralogy and Petrology, *12*, 202-222.

Krylov, A. Ya. and Atrashenok, L.Ya., 1959, The mode of occurrence of uranium in granites: Geochemistry, no. 3, 307-313.

Larsen, E.S., 3rd., 1957, Distribution of uranium in igneous complexes: U.S. Atomic Energy Comm. Rept. TEI-700, 249-253.

Larsen, E.S., Jr. and Gottfried, D., 1961, Distribution of uranium in rocks and minerals of Mesozoic batholiths in western United States: U.S. Geol. Survey Bull. 1070-C, 63-103.

Larsen, E.S., Jr., Phair, G., Gottfried, D. and Smith, W.S., 1956, Uranium in magmatic differentiation: Internat. Conf. Peaceful Uses of Atomic Energy, *6*, 240-247.

Le, V.T. and Stussi, J.M., 1973, Les minéraux d'uranium et de thorium des granites de la Montagne Bourbonnaise (Massif Central français): Sci. Terre, *18*, 353-379.

Leonova, L.L. and Tauson, L.V., 1958, The distribution of uranium in the minerals of Caledonian granitoids of the Susamyr batholith (central Tien Shan): Geochemistry, no. 7, 815-826.

Lyons, J.B., 1964, Distribution of thorium and uranium in three early Paleozoic plutonic series of New Hampshire: U.S. Geol. Survey Bull. 1144-F, 43 pp.

Lyons, J.B. and Larsen, E.S., 3rd., 1958, Uranium and thorium content of three early Paleozoic plutonic series in New Hampshire: U.S. Atomic Energy Comm. Rept. TEI-750, 108-112.

Magne, R., Berthelin, J. and Dommergues, Y., 1975, Solubilisation et insolubilisation de l'uranium des granites par des bactéries hétérotrophes: in *Formation of Uranium Ore Deposits*, 73-86, Internat. Atomic Energy Agency, Vienna.

Malan, R.C., 1972, Summary report—Distribution of uranium and thorium in the Precambrian of the western United States: U.S. Atomic Energy Comm. Rept. AEC-RD-12, 59 pp.

Malan, R.C. and Sterling, D.A., 1970, Distribution of uranium and thorium in the Precambrian of the west-central and northwest United States: U.S. Atomic Energy Comm. Rept. AEC-RD-11, 64 pp.

Marjaniemi, D.K. and Basler, A.L., 1972, Geochemical investigations of plutonic rocks in the western United States for the purpose of determining favorability for vein-type uranium deposits: U.S. Atomic Energy Comm. Rept. GJO-912-16, 134 pp.

Neuerburg, G.J., 1956, Uranium in igneous rocks of the United States: U.S. Geol. Survey Prof. Paper 300, 55-64.

Page, L.R., 1960, The source of uranium in ore deposits: 21st Internat. Geol. Cong., pt. 15, 149-164.

Phair, G., 1952, Radioactive Tertiary porphyries in the Central City district, Colorado, and their bearing upon pitchblende deposition: U.S. Atomic Energy Comm. Rept. TEI-247, 53 pp.

Phair, G., 1958, Uranium and thorium in the Laramide intrusives of the Colorado Front Range: U.S. Atomic Energy Comm. Rept. TEI-750, 103-108.

Phair, G., 1975, Uranium, thorium, and lead in igneous rocks and veins, Colorado: U.S. Geol. Survey Open-file Rept. 75-501.

Phair, G. and Gottfried, D., 1964, The Colorado Front Range, Colorado, U.S.A., as a uranium and thorium province: in Adams, J.A.S. and Lowder, W.M., eds., *The Natural Radiation Environment,* 7-38, University of Chicago Press.

Pliler, R. and Adams, J.A.S., 1962, The distribution of thorium and uranium in a Pennsylvanian weathering profile: Geochim. et Cosmochim. Acta, *26,* 1137-1146.

Ranchin, G., 1968, Contribution à l'étude de la répartition de l'uranium à l'état de traces dans les roches granitiques saines: Sci. Terr, *13,* 159-205.

Ranchin, G., 1970, La géochimie de l'uranium et la différenciation granitique dans la province du nord-Limousin: Sci. Terre Mém. 17, 483 pp.

Renard, J.P., 1970, Géochimie de l'uranium de surface dans les massifs granitiques vendéens: Sci. Terre, *15,* 167-194.

Richardson, K.A., 1964, Thorium, uranium, and potassium in the Conway Granite, New Hampshire, U.S.A.: in Adams, J.A.S. and Lowder, W.M., eds., *The Natural Radiation Environment,* 39-50, University of Chicago Press.

Rogers, J.J.W., 1964, Statistical tests of the homogeneity of the radioactive components of granitic rocks: in Adams, J.A.S. and Lowder, W.M., eds., *The Natural Radiation Environment,* 51-62, University of Chicago Press.

Rogers, J.J.W. and Ragland, P.C., 1961, Variation of thorium and uranium in selected granitic rocks: Geochim. et Cosmochim. Acta, *25,* 99-109.

Rogers, J.J.W., Adams, J.A.S. and Gatlin, B., 1965, Distribution of thorium, uranium, and potassium concentrations in three cores from the Conway granite, New Hampshire, U.S.A.: Am. Jour. Sci., *263,* 817-822.

Rosholt, J.N. and Noble, D.C., 1969, Loss of uranium from crystallized silicic volcanic rocks: Earth and Planetary Sci. Letters, *6*, 268-270.

Rosholt, J.N., Prijana and Noble, D.C., 1971, Mobility of uranium and thorium in glassy and crystallized volcanic rocks: Econ. Geology, *66*, 1061-1069.

Sighinolfi, G.P. and Sakai, T., 1974, Uranium and thorium in potash-rich rhyolites from western Bahia (Brazil): Chem. Geology, *14,* 23-30.

Smith, D.L., Garvey, M.J. and Davis, M.P., 1975, Uranium, thorium, and potassium abundances in rocks of the Piedmont of Georgia (abst.): EOS (Am. Geophys. Union Trans.), *56,* 467.

Stuckless, J.S., 1975, Uranium mobility, Granite Mountains, Wyoming (abst.): U.S. Geol. Survey Open-file Rept. 75-595, 49.

Szalay, A. and Samsoni, Z., 1973, Investigations of the leaching of uranium from crushed magmatic rocks: in Ingerson, E., ed., *Proceedings of Symposium on Hydrogeochemistry and Biogeochemistry, 1,* 261-272.

Vollmer, R., 1976, Rb.-Sr. and U-Th-Pb systematics of alkaline rocks: the alkaline rocks from Italy: Geochim. et Cosmochim. Acta, *40*, 283-295.

Wells, J.D. and Harrison, J.E., 1961, Petrography of radioactive Tertiary igneous rocks, Front Range mineral belt, Colorado: U.S. Geol. Survey Bull. 1032-E, 223-272.

Whitfield, J.M., Rogers, J.J.W. and Adams, J.A.S., 1959, The relationship between the petrology of the thorium and uranium contents of some granitic rocks: Geochim. et Cosmochim. Acta, *17,* 248-271.

Zielinski, R.A., 1975, Uranium in rhyolitic lavas (abst.): U.S. Geol. Survey Open-file Rept. 75-595, 54.

Appendix II. References to data on the uranium content of ground waters.

Alekseev, F.A., Gottikh, R.P., Soifer, V.N. and Brezgunov, V.S., 1970, Radioactive elements and deuterium in formation waters of the Bukhara-Karshin artesian basin: Geochem. International, *7,* 1030-1039.

Alekseev, F.A., Yermakov, V.I. and Filonov, V.I., 1958, Radioactive elements in oil field waters: Geochemistry, no. 7, 806-814.

Barker, F.B. and Scott, R.C., 1958, Uranium and radium in the groundwater of the Llano Estacado, Texas and New Mexico: Am. Geophys. Union Trans., *39,* 459-466.

Barker. F.B. and Scott, R.C., 1961, Uranium and radium in groundwater from igneous terranes of the Pacific Northwest: U.S. Geol. Survey Prof. Paper 424-B, 298-299.

Butler, A.P., 1969, Groundwater as related to the origin and search for uranium deposits in sandstone: Contr. Geology, *8*, no. 2, pt. 1, 81-86.

Cohen, P., 1961, An evaluation of uranium as a tool for studying the hydrogeochemistry of the Truckee Meadows area, Nevada: Jour. Geophys. Research, *66,* 4199-4206.

Cowart, J.B., 1974, U^{234} and U^{238} in the Carrizo Sandstone aquifer of south Texas: unpub. doctoral thesis, Florida State University, 80 pp.

Dall 'Aglio, M., 1972, Geochemical exploration for uranium: in *Uranium Exploration Methods,* 189-208, Internat. Atomic Energy Agency, Vienna.

Denson, N., Zeller, H. and Stephens, J., 1956, Water sampling as a guide in the search for uranium deposits and its use in evaluating widespread volcanic units as potential source beds for uranium: U.S. Geol. Survey Prof. Paper 300, 673-680.

Doi, K., Hirono, S. and Sakamaki, Y., 1975, Uranium mineralization by groundwater in sedimentary rocks, Japan: Econ. Geology, *70*, 628-646.

Fix, P.F., 1956, Hydrogeochemical exploration for uranium: U.S. Geol. Survey Prof. Paper 300, 667-671.

Germanov, A.I., Batulin, S.G., Volkov, G.A., Lisitsin, A.K. and Serebrennikov, V.S., 1958, Some regularities of uranium distribution in underground waters: Internat. Conf. Peaceful Uses of Atomic Energy, *2*, 161-177, United Nations.

Hecht, F., Keupper, H. and Petrascheck, W.E., 1958, Preliminary remarks on the determination of uranium in Austrian springs and rocks: Internat. Conf. Peaceful Uses of Atomic Energy, *2,* 158-160, United Nations.

Judson, S. and Osmond, J.K., 1955, Radioactivity in ground and surface water: Am. Jour. Sci., *253,* 104-116.

Jurain, G., 1957, Remarques sur la teneur à uranium des eaux des Vosges méridionales: Acad. Sci. Comptes Rendus, Ser. D, *245,* 1071-1074.

Kuptsov, V.M. and Cherdyntsev, V.V., 1969, The decay products of uranium and thorium in active volcanism in the U.S.S.R.: Geochem. International, *6,* 532-545.

Kuroda, K., 1944, Strongly radioactive springs discovered in Masutomi: Bull. Chem. Soc. Japan, *19,* no. 3, 33-83.

Kyuregyan, T.N. and Kocharyan, A.G., 1969, Migration forms of uranium in carbonate waters of a Caucasian district: Internat. Geology Rev. *11,* 1087-1089.

Landis, E.R., 1971, Uranium content of ground and surface waters in a part of the central Great Plains: U.S. Geol. Survey Bull. 1087-G, 223-258.

Lisitsin, A.K., 1971, Ratio of the redox equilibria of uranium and iron in strataform aquifers: Internat. Geology Rev., *13,* 744-751.

Lopatkina, A.P., 1964, Characteristics of migration of uranium in natural waters of humid regions and their use in the determination of the geochemical background for uranium: Geochem. International, no. 4, 788-795.

Murakami, Y., Fugiwara, S., Sato, M. and Ohashi, S., 1958, Internat. Conf. Peaceful Uses of Atomic Energy, *2,* 131-139.

Osmond, J.K., 1964, The distribution of the heavy radioelements in the rocks and waters of Florida: in Adams, J.A.S. and Lowder, W.M., eds., *The Natural Radiation Environment,* 153-159, University of Chicago Press.

Osmond, J.K. and Cowart, J.B., 1974, U^{234}/U^{238} variations in a sandstone aquifer (abs): Trans. Am. Geophys. Union, *55,* 458.

Osmond, J.K., Kaufman, M.I. and Cowart, J.B., 1974, Mixing volume calculations, sources, and aging trends of Floridian aquifer water by uranium isotopic methods: Geochim. et Cosmochim. Acta, *38,* 1083-1100.

Paces, T., 1969, Chemical equilibria and zoning of subsurface water from Jáchymov ore deposit, Czechoslovakia: Geochim. et Cosmochim. Acta, *33,* 591-609.

Phoenix, D.A., 1959, Occurrence and chemical character of groundwater in the Morrison Formation: U.S. Geol. Survey Prof. Paper 320, 55-64.

Rogers, J.J.W. and Adams, J.A.S., 1967, Uranium: in Wedepohl, K.H., ed., *Handbook of Geochemistry, 2,* pt. 1, Chapter 92, 50 pp., Springer-Verlag, Berlin.

Ross, J.R. and George, D.R., 1971, Recovery of uranium from natural mine waters by counter current ion exchange: U.S. Bur. Mines Rept. Inv. 7471, 17 pp.

Scott, R.C. and Barker, F.B., 1962, Data on uranium and radium in groundwater in the United States 1954-1957: U.S. Geol. Survey Prof. Paper 426, 115 pp.

Smith, G.H. and Chandler, T.R.D., 1958, A field method for the determination of uranium in natural waters: Internat. Conf. Peaceful Uses of Atomic Energy, *2,* 148-152, United Nations.

Souther, J.G., 1974, Geothermal project: Canada Geol. Survey Paper 74-1A, 41-42.

Titayeva, N.A., Filonov, V.A., Ovchenkov, V. Ya., Veksler, T.I., Orlova, A.V. and Tyrina, A.S., 1973, Behavior of uranium and thorium isotopes in crystalline rocks and surface waters in a cold, wet climate: Geochem. International, *10,* 1146-1151.

Udaltsova, N.I. and Leonova, L.L., 1970, Uranium, thorium and rare earth elements in thermal waters of the Kamchatka Peninsula: Geokhmiya, no. 12, 1504-1510.

White, D.E., Hem, J.D. and Waring, G.A., 1963, Chemical composition of subsurface waters: U.S. Geol. Survey Prof. Paper 440-F, 67pp.

AUTHOR INDEX

K

L

M

R

S

T

U

V

SUBJECT INDEX

C

D

H

J

K

L

M

N

O

T

V